T0320977

Wind Energy

AN INTRODUCTION

Wind Energy

AN INTRODUCTION

Mohamed A. El-Sharkawi

UNIVERSITY OF WASHINGTON, SEATTLE, USA

CRC Press
Taylor & Francis Group
Boca Raton   London   New York

CRC Press is an imprint of the
Taylor & Francis Group, an **informa** business

CRC Press
Taylor & Francis Group
6000 Broken Sound Parkway NW, Suite 300
Boca Raton, FL 33487-2742

© 2016 by Taylor & Francis Group, LLC
CRC Press is an imprint of Taylor & Francis Group, an Informa business

No claim to original U.S. Government works

Printed and bound in India by Replika Press Pvt. Ltd.

Printed on acid-free paper
Version Date: 20141017

International Standard Book Number-13: 978-1-4822-6399-2 (Hardback)

Visit the Taylor & Francis Web site at
http://www.taylorandfrancis.com

and the CRC Press Web site at
http://www.crcpress.com

To wind energy researchers and developers for their visions and tenacity to pursue

the technology even when renewable energy was unrealistically expensive

To my wife Fatma and sons Adam and Tamer

Contents

Preface

Wind energy has become an important source of electricity worldwide. Wind power plants are installed with high capacities all over the world. Their penetration ratio can often exceed 10% in several areas in the United States and Europe. With modern designs and control, wind power plants are now comparable to conventional generations in terms of capacity. This technology is ready to move from research to educational curricula. Indeed, students in electric power engineering need to be versed in this technology to meet the industry requirements for future renewable energy specialists.

One of the major challenges for achieving this objective is that most wind energy books are research- or industrial-oriented, which could be hard to adopt in the university curricula. This book is addressing the need for an undergraduate/graduate textbook in wind energy which comprehensively covers the main aspects of wind energy types, operation, modeling, analysis, integration, and control. This book has lots of modeling, examples, and exercise problems, which are key elements for university education. Some of the examples are from real events.

The background needed for this book is the basic electric circuit theory. Thus, it is suitable for students at the senior or graduate levels. In addition, the book is written for practicing engineers.

In Chapter 1, the history of the development of wind energy system is provided. In Chapter 2, the aerodynamic theories that govern the operation of wind turbines are explained with enough illustrations to help students with no background in this area understand the concept. In addition, the separation of wind turbine at the farm site is addressed and evaluated.

In Chapter 3, wind energy statistics are covered to address the stochastic nature of wind speed. The modeling of wind speed as a probability density function is used to evaluate sites for wind energy generation.

In Chapter 4, the common types of wind turbines are given. The differences between the types are highlighted. In Chapter 5, the main power electronic circuits used in wind energy are explained, modeled, and analyzed. In Chapter 6, the induction generator is discussed in detail from the basic principle of induced voltage to the steady-state and dynamic models. In Chapter 7, the salient pole, cylindrical rotor, and permanent magnet synchronous generator types are discussed. In addition, the steady-state and dynamic model are derived.

In Chapter 8, type 1 wind turbine is analyzed in detail in terms of operation, stability, control, and protection. In Chapter 9, type 2 is similarly analyzed. In Chapter 10, type 3, which is the most common type and the hardest to analyze, is presented in gradual steps to make it easier for an average undergraduate student to understand. The operation and protection of this doubly fed induction generator is presented and evaluated. In Chapter 11, the operation, control, and protection of type 4 turbine is given.

Chapter 12 is dedicated to the main integration challenges of wind energy systems with electric utility systems. The chapter analyzes key integration issues from the wind farm and utility viewpoints. These include system stability, fault ride-through, variability of wind speed, and reactive power. The chapter also provides methods by which successful integration can be achieved.

Mohamed A. El-Sharkawi
University of Washington

Author

Mohamed A. El-Sharkawi is a fellow of the IEEE. He received his undergraduate education from Helwan University in Egypt in 1971 and his PhD from the University of British Columbia in 1980. He joined the University of Washington as a faculty member in 1980. He is presently a professor of electrical engineering in the energy area. He has also served as the associate chair and the chairman of graduate studies and research. Professor El-Sharkawi served as the vice president for technical activities of the IEEE Computational Intelligence Society and is the founding chairman of the IEEE Power and Energy Society's subcommittee on renewable energy machines and systems. He is the founder and cofounder of several international conferences and the founding chairman of numerous IEEE task forces, working groups, and subcommittees. He has organized and chaired numerous panels and special sessions in IEEE and other professional organizations. He has also organized and taught several international tutorials on power systems, renewable energy, electric safety, induction voltage, and intelligent systems. Professor El-Sharkawi is an associate editor and a member of the editorial boards of several engineering journals. He has published over 250 papers and book chapters in his research areas. He has authored three textbooks—*Fundamentals of Electric Drives*, *Electric Energy: An Introduction*, and *Electric Safety: Practice and Standards*. He has also authored and coauthored five research books in the area of intelligent systems and power systems. He holds five licensed patents in the area of renewable energy, VAR management, and minimum arc sequential circuit breaker switching.

For more information, please visit El-Sharkawi's website at http://cialab.ee.washington.edu.

List of Variables

A	Area
A_d	Cross-sectional area of the air mass at far downstream distance
A_{blade}	Cross-sectional area of the air mass through the turbine blades
A_u	Cross-sectional area of the air mass at far upstream distance
B	Flux density
C_D	Drag coefficient
C_L	Lift coefficient
C_p	Coefficient of performance
d	Diameter
E_{a2}	Induced voltage in rotor at standstill
E_{co}	Stored energy in the capacitor during steady-state operation
E_f	Equivalent field voltage
E_f'	Voltage behind transient reactance
e_{fd}	Steady-state equivalent field voltage
E_r	Induced voltage in rotor while spinning
$E[w]$	Expected value of wind speed
f	Frequency in ac circuits
F	Aerodynamic force
F_D	Force of drag
F_L	Force of lift
f_s	Frequency of a reference signal
F_{xy}	Cumulative distribution function (CDF)
g	Gravitational acceleration (9.8 m/s^2)
h	Height
H	Head speed or inertia constant of the rotating mass
i	Instantaneous current
I_{a2}	Rotor current
I_{a2}'	Rotor current referred to the stator winding
I_{ave}	Average value of a current waveform
I_d	Instantaneous direct axis current
i_{f-max}	Maximum fault current
I_{max}	Maximum (peak) value of current waveform
I_m	Current in magnetizing branch
i_o	Instantaneous zero sequence current
i_{on}	Current during the on time of a switch
I_q	Instantaneous quadrature axis current
I_{rms}	Root mean square value of current waveform
J	Moment of inertia
K	Duty ratio in power electronics, or scale factor in electric machine
KE	Kinetic energy
L	Inductance
L_d	Direct axis inductance
L_m	Mutual inductance

(Continued)

L_q	Quadrature axis inductance
ℓ_{xx}	Self-inductance of coil x
ℓ_{xy}	Mutual inductance between coil x and y
m	Mass in mechanical terms or modulation index (or duty ratio) in power electronics
M_w	Molecular weight
N_1 and N_2	Number of turns in transformer windings
N_h and N_l	Number of turns in high- and low-voltage windings
n_s	Synchronous speed
P	Power or number of poles or real power
P_{blade}	Power captured by the blade
P_{cu}	Copper loss of windings
P_{core}	Core loss
P_d	Developed power
P_{df}	Forecasted demand
P_g	Airgap power in induction machine or grid power in type 4 system
P_{in}	Input power
P_{out}	Output power
pp	Number of pole pairs
P_r	Pressure
P_r	Rotor real power
P_{RDR}	Power consumed by rotor dynamic resistance
P_s	Stator real power
P_{SDR}	Power consumed by series dynamic resistance
P_{slip}	Slip real power
P_w	Wind power
P_{wf}	Forecasted wind power
Q	Reactive power
Q_{fcb}	Reactive power at the farm collection bus
Q_g	Reactive power at the grid bus
Q_{gsc}	Reactive power produced by the grid-side converter
Q_{out}	Output reactive power
Q_r	Rotor reactive power
Q_{rs}	Reactive power produced in the rotor circuit due to the injected voltage and the slip of the machine
Q_s	Stator reactive power
Q_{tl}	Reactive power consumed by transmission line
R	Ideal gas constant in aerodynamics and resistance in electric circuits
r_1	Resistance of the stator windings
r_2	Resistance of the rotor windings
r_2'	Resistance of the rotor windings referred to stator
r_a	Armature resistance
r_{add}	Inserted resistance in rotor circuit
R_b	Dynamic braking resistance
r_c	Distance from hub to center of gravity of blade
R_{cb}	Crowbar resistance
R_{ch}	Chopper resistance
r_d	Developed resistance
r_f	Resistance of field winding

(Continued)

R_{RDR}	Rotor dynamic resistance
R_{SDR}	Series dynamic resistance
r_{th}	Thevinin's equivalent resistance
s	Slip
S	Separation in aerodynamics, or apparent power in electric circuits
s'	Slip at maximum power
s^*	Slip at maximum torque
t	Time
T	Torque in machines, or temperature in solid state and aerodynamics
T_{blade}	Torque of the blades
T_d	Developed electric torque
T_e	Electric torque
T_m	Mechanical torque
t_{off}	Off time of a switch
t_{on}	On time of a switch
TSR	Tip-speed ratio
u_x	Unit step function, its value is zero unless $\omega t \geq x$
V_{a2}	Injected voltage in rotor circuit
V'_{a2}	Injected voltage in rotor circuit referred to the stator winding
v_{ar}	Reference voltage of phase a
Var	Variance
V_{ave}	Average value of a voltage waveform
v_{car}	Carrier voltage signal
v_d	Instantaneous direct axis voltage
V_d	rms direct axis voltage
V_D	Developed voltage
V_{dc}	dc voltage
V_{DVR}	Output voltage of dynamic voltage regulator
V_g	Grid voltage
V_i	Injected voltage
V_{max}	Maximum (peak) value of voltage waveform
v_o	Instantaneous zero sequence voltage
V_o	rms zero sequence voltage
vol	Volume
V_{poi}	Voltage at point of interconnection
v_q	Instantaneous quadrature axis voltage
V_q	rms quadrature axis voltage
V_{rms}	Root mean square value of voltage waveform
v_s	Instantaneous voltage source
v_t	Instantaneous terminal voltage across load
V_{th}	Thevinin's equivalent voltage
V_{xn}	rms voltage between point x and the neutral
V_{xy}	rms voltage between points x and y
w	Wind speed
w_d	Downstream wind speed
w_r	Relative wind speed
w_u	Upstream wind speed
x_1	Inductive reactance of the stator windings

(Continued)

x_2	Inductive reactance of the rotor windings
x_2'	Inductive reactance of the rotor windings referred to stator
x_d	Direct axis inductive reactance
x_f	Inductive reactance of field winding
x_m	Inductive reactance of the magnetizing branch
x_q	Quadrature axis inductive reactance
x_r	Inductive reactance of rotor at rotor frequency
x_s	Synchronous reactance
x_{th}	Thevinin's equivalent inductive reactance
z	Impedance in electric circuits
z_{th}	Thevinin's equivalent impedance

List of Symbols

α	Angle of attack in aerodynamics or triggering angle in power electronics
β	Pitch angle in aerodynamics or commutation angle in power electronics
γ	Conduction period in power electronics
$\Gamma(\cdot)$	Gama function
δ	Density of material or power angle
ε	Error
ε_d	Error in forecasted demand
ε_w	Error in forecasted wind power
η	Efficiency
θ	Power factor angle
θ_2	Angle between the rotor and stator axes of induction motor
θ_s	Angle between the stator and direct axes
λ	Tip-speed ratio in aerodynamics or flux linkage in electric circuits
λ_d	Direct axis flux
λ_q	Quadrature axis flux
λ_o	Zero sequence flux
μ	Specific gas constant
ξ	Damping coefficient
ρ	Power density in aerodynamics or instantaneous power in electric circuits
ρ_{xy}	Cross-correlation coefficient between samples x and y
σ	Standard deviation
T	Time constant of the load
τ_{do}'	Open circuit field voltage
Φ	Flux

(*Continued*)

ω	Angular speed in mechanical terms or angular electrical frequency in electrical terms
ω_{blade}	Angular speed of blades
ω_n	Natural frequency of oscillation
ω_s	Angular synchronous speed
$\mathfrak{R}_d$	Reluctance seen by the flux crossing the airgap through the direct axis
$\mathfrak{R}_q$	Reluctance seen by the flux crossing the airgap through the quadrature axis

1

History of the Wind Energy Development

Wind has been a source of energy all along history. The ancient Egyptians discovered the power of wind, which led to the invention of sailboats around 5000 BC. Although no one knows exactly who invented the first windmill, archaeologists discovered a Chinese vase dating back to the third millennium BC that had an image resembling a windmill. By 200 BC, the Persian, Chinese, and Middle Easterners used windmills extensively for irrigation, wood cutting, and grinding grains. They were often constructed as revolving door systems with woven reed sails, similar to the vertical-axis wind system used today. By the eleventh century, people in the Middle East were using windmills extensively for food production. During the period from the eleventh century to thirteenth century, foreign merchants who traded with the Middle East, and the crusaders who invaded the region, carried the windmill technology back to Europe. Figure 1.1 shows a nineteenth-century renovated windmill in Europe. In Holland, windmills were also used to drain lands below the water level of the Rhine River. During this era, working in windmills was one of the most hazardous jobs in Europe. The workers were frequently injured because windmills were constructed of a huge rotating mass with little or no control on its rotation. The grinding or hammering sounds were so loud that many workers became deaf, the grinding dust of certain material such as wood caused respiratory health problems, and the grinding stones often caused sparks and fires.

In addition to producing mechanical power, windmills were used to communicate with neighbors by locking the windmill sails in a certain arrangement. During World War II, the Netherlanders used to set windmill sails in certain positions to alert the public of a possible attack by their enemies.

During the nineteenth century, the European settlers brought windmill technology to North America. They were mainly used to pump water from wells for farming. The first known windmill was built by Daniel Halladay in 1854. It was quite an innovative system, as it was able to align itself with wind direction. In 1863, he established the U.S. Wind Engine & Pump Company, Illinois, which was the first mass manufacturer of windmills in the United States. One of their designs is shown in Figure 1.2. During the nineteenth and early twentieth centuries, there were over 1000 factories building these very useful machines. Most, however, were weak designs that break due to over speeding during wind gusts.

Windmills were initially made out of wood, which limited their powers and speeds. Over time, iron and steel replaced wood, and systems with gearbox were introduced. They were much powerful systems, but much more expensive than wood. The first all-steel windmill was invented and designed by Thomas Osborn Perry in 1883.

In 1888, Charles Francis Brush of the United States made a major innovation by converting the kinetic energy in wind into electrical energy. These types of windmills are called "wind turbines." The first design, which is shown in Figure 1.3, was about 20 m in height and 36 ton in weight. This enormous structure produced just 12 kW. Because power grid did not reach farmlands in the United States until the second quarter of the twentieth century, farmers relied on these wind turbines for their electric energy needs. During the period from 1930 to 1940, thousands of wind turbines were used in rural areas not yet served

FIGURE 1.1
Renovated nineteenth-century windmill.

FIGURE 1.2
Halladay's windmill. (Courtesy of Billy Hathorn through Wikipedia.)

FIGURE 1.3
First electric wind turbine. (Courtesy of Robert W. Righter through Wikipedia.)

by the power grid. The Great Plains (west of the Mississippi River and east of the Rocky Mountains in the United States and Canada) had the majority of these machines.

In 1891, Poul La Cour of Denmark built the first wind turbine outside of the United States. In 1896, he tested small models of wind turbines in a wind tunnel. This was the first of such experiments in the world. Among his major contributions is the discovery of the power-capturing capability as a function of blade shape and number of blades. His primitive experiment in wind tunnel showed that eight blades can capture about 28% of the available wind energy, whereas 16 blades can capture about 29%. La Cour concluded that the number of blades and the energy-capturing capability are not linearly related. In addition, he showed that curved blades could capture more energy from wind. These are key factors that resulted in the designs of current wind turbines.

After the invention of the steam engine and the expansion of power grids to rural areas, interest in wind turbines declined. The interest was only renewed during the oil crisis of the 1970s, mainly because the generous tax credits by the U.S. government. Consequently, several wind farms were built in the United States in the 1970s and 1980s. These wind turbines, unfortunately, were very expensive and high-maintenance machines. They also created electrical problems to the grid such as voltage flickers and voltage depression due to the high and cyclic demand for reactive powers.

Interest in wind energy declined again in the 1980s because of the following four reasons:

1. Oil prices dropped substantially around 1985.
2. U.S. tax credits were provided for anyone who had installed wind turbines instead of the actual energy production. Because of this shortcoming, wind turbines were afflicted with low productivity and frequent failures. It was not unusual

to find a wind farm with less than 10% of their turbines producing electricity. This investment tax credits expired in 1986.

3. Designs of wind turbines were fragile and required extensive maintenance.
4. Cost of electricity generated by wind turbines were several times higher than those provided from conventional resources.

To address the declining interest in wind energy, the United States issued a new type of tax credit in 1992 based on the production of electricity rather than cost of installation, known as federal production tax credit (PTC). PTC encouraged major improvements in wind turbine research and designs, and encouraged developers to maximize their electricity production. As a result, nowadays, the cost of wind energy dropped to a level comparable to fossil-fuel power plants.

1.1 Wind Turbines

Modern wind turbines are much larger in size and much more reliable than the 1970s–1980s versions. The power rating of wind turbines, as shown in Figure 1.4, has increased from just a few kilowatt to up to 8 MW for a single unit in 2013. Because the air density is low, these machines are large in size, as seen in the figure. Keep in mind that the height of the Statue of Liberty is 93 m and that of the Great Pyramid is 140 m.

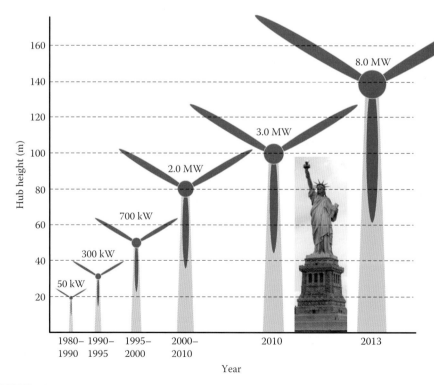

FIGURE 1.4
Average height of wind turbines.

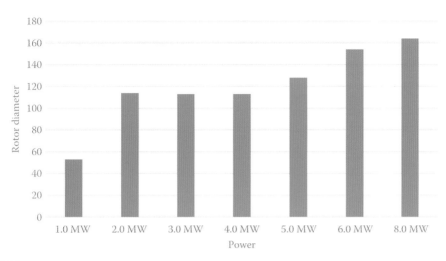

FIGURE 1.5
Typical rotor diameter.

The power captured by the turbine is proportional to the sweep area of its blades. This makes the power proportional to the square of the blade length, as seen in Chapter 2. The diameter of the sweep area is known as the "rotor diameter," which is twice the length of a single blade. Some typical rotor diameters is given in Figure 1.5. To put the number into perspective, the diameter of a 2 MW turbine is more than the length of a Boeing 747 airplane or an Airbus 380.

Figure 1.6 shows a 1.8 MW turbine blade. Note the length of the blade with respect to the extended-load truck. Such a large length poses a transportation problem as most roads cannot allow drivers to negotiate turns. This is why larger turbines are built offshore.

The drive shaft of wind turbines can rotate horizontally or vertically. A horizontal-axis wind turbine (HAWT) is shown in Figure 1.7. This is the most common type of wind turbine system used today. Its main drive shaft, gearbox, electrical generator, and, sometimes, the transformer are housed in the nacelle at the top of a tower (see Figure 1.8). The turbine is aligned to face the upwind. To prevent the blades from hitting the tower at high wind conditions, the blades are placed at a distance in front of the tower and tilted up a little. The tall tower allows the turbine to access strong wind. Every blade receives power from wind at any position, which makes the HAWT a high-efficient design. The HAWT, however, requires

FIGURE 1.6
Blade of a 1.8 MW wind turbine.

FIGURE 1.7
A 1.8 MW horizontal-axis wind turbine.

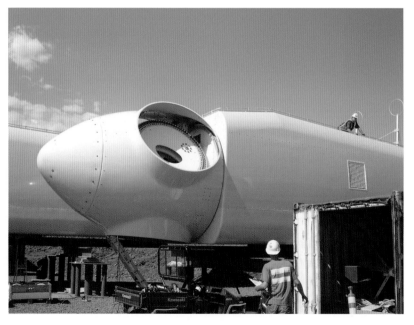

FIGURE 1.8
Nacelle of a 1.8 MW wind turbine.

massive tower construction to support the heavy nacelle, and it requires an additional yaw control system to turn the blades toward wind.

The other design is the vertical-axis wind turbine (VAWT) shown in Figure 1.9. It is known as "Darrieus wind turbines" and it looks like a giant upside down eggbeater. The VAWT was among the early designs of wind turbines because it is suitable for sites with shifting wind directions. This design does not require a yaw mechanism to direct the blade into wind. The generator, gearbox, and transformers are all located at the ground level, making the VAWT easier to install and maintain as compared with the HAWT. The cut-in speed of the VAWT is generally lower than that for the HAWT. However, because of its massive inertia, VAWT may require external power source to startup the turbine, and extensive bearing system to support the heavy weight of the turbine. Because wind speed is slower near ground, the available wind power is lower than that of HAWT. In addition, objects near ground can create turbulent flow that can produce vibration on the rotating components and cause extra stress on the turbine.

The VAWT is also popular in small wind energy systems. One of them is shown in Figure 1.10. This small VAWT is intended for individual use (home or office), and several units with design variations are installed all over the world.

FIGURE 1.9
Vertical-axis wind turbine. (Courtesy of U.S. National Renewable Energy Lab.)

FIGURE 1.10
Small wind turbine. (Courtesy of Anders Sandberg through Wikipedia.)

1.2 Offshore Wind Turbines

With the continuous demand for larger wind turbines, researchers envisioned the offshore wind turbines. This is because of several reasons; a few among them are as follows:

- Size of wind turbines will eventually reach a level where roads cannot accommodate the transportation of the blades.
- Offshore wind is stronger than onshore.
- Offshore winds are often strong in the afternoon, which match the time of heavy electricity demand.
- Most densely populated areas are near shores. Thus, offshore systems do not need extensive transmission systems. For example, 28 states in the United States have coastal lines. These states consume 78% of the national electric energy.
- Offshore turbines are not normally visible from shores. This reduces the public concern with regard to the visual impact of wind farms.
- Noise and light flickers are less of a problem for offshore turbines.

FIGURE 1.11
Offshore vertical-axis wind turbine. (Courtesy of Leonard G. through Wikipedia.)

With today's technology, most of the offshore installations are in relatively shallow water (up to 50 m deep) (Figure 1.11). The first offshore wind turbines were installed in Denmark in 1991. By early 2014, 70 offshore wind farms were in operation with a capacity of about 7 GW. Offshore wind is expected to dominate the large turbine market for the foreseeable future.

Exercise

1. What is the difference between a wind mill and a wind turbine?
2. What is a Halladay windmill?
3. Who invented the first wind turbine?
4. Where was the first wind turbine invented?
5. State one of the major contributions of Poul La Cour.
6. What is the average blade length for a 6 MW wind turbine?
7. What are the advantages and disadvantages of HAWT?
8. What are the advantages and disadvantages of VAWT?
9. What are the advantages and disadvantages of offshore wind turbines?

2

Aerodynamics of Wind Turbines

The role of wind turbines (WTs) is to harness the kinetic energy in wind and convert it into electrical energy. According to Newton's second law of motion, the kinetic energy of an object is the energy it possesses while in motion.

$$KE = \frac{1}{2}mw^2 \tag{2.1}$$

where:
  KE is the kinetic energy of the moving object (Watt second, Ws)
  m is the mass of the object (kg)
  w is the velocity of the object (m/s)

If the moving object is air, KE of the moving air (wind) can be computed in a similar way. In Figure 2.1, the mass of air passing through a ring is

$$m = \text{vol } \delta \tag{2.2}$$

where:
  δ is the density of air (for thin air, we can use 1.0 kg/m³)
  vol is the volume of air passing through the ring

The volume of air passing through the ring is the area of the ring multiplied by the length of the air column.

$$\text{vol} = Ad \tag{2.3}$$

where:
  A is the area of the ring
  d is the length of the air column, which changes with time. It depends on the velocity of wind and time

$$d = wt \tag{2.4}$$

where:
  t is time (s)
  w is wind speed (m/s)

Hence, the mass of air passing through the ring during a given time is

$$m = Aw\delta t \tag{2.5}$$

Substituting the mass in Equation 2.5 into Equation 2.1 yields

$$KE = \frac{1}{2}A\delta t w^3 \tag{2.6}$$

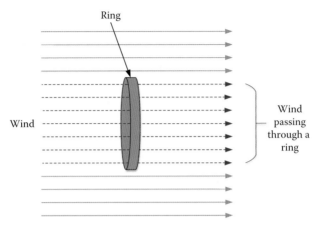

FIGURE 2.1
Wind passing through a ring.

Because the energy is power multiplied by time, the wind power (P_w) in watt is

$$P_w = \frac{KE}{t} = \frac{1}{2} A \delta w^3 \tag{2.7}$$

Note that the KE and the power of wind are proportional to the cube of the speed of wind; if the wind speed increases by just 10%, the KE of wind increases by 33.1%.

From Equation 2.7, the wind power density can be written as

$$\rho = \frac{P_w}{A} = \frac{1}{2} \delta w^3 \tag{2.8}$$

For a dry thin air of 1 kg/m³, the wind power density is about 3.0 kW/m² if wind speed is 18 m/s. This is a tremendous amount of energy for moderate wind speeds. This is why storms are destructive; at 35 m/s (78 mile/hr), wind power density is about 21.5 kW/m². Figure 2.2 shows the wind power density for various weak to moderate wind speeds.

Wind power density is often used to evaluate the potentials of sites for electric energy production. Figure 2.3 shows the map of average wind speed in the United States at an

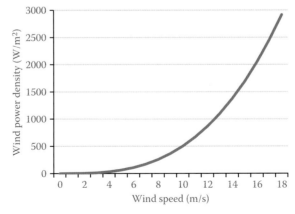

FIGURE 2.2
Wind power density as a function of wind speed.

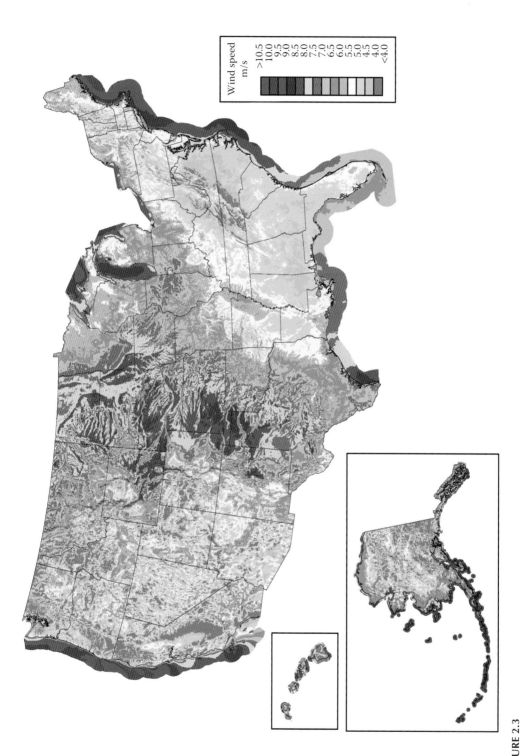

FIGURE 2.3
The average wind power density map of the United States at 80 m above sea level. (Courtesy of the US National Renewable Energy Laboratory.)

elevation of 80 m. Several offshore areas in the east and west coasts as well as the Great Lakes region have fresh to strong breezes with an average wind speed of about 10.0 m/s. These sites have an average annual power density of about 500 W/m².

2.1 Wind Speed

Wind is a renewable resource as it is in constant motion relative to the Earth's surface. Wind speed is a stochastic variable; its magnitude and direction are continuously changing and cannot be controlled. The three main factors that determine wind speed are pressure gradient force (PGF), Coriolis force, and friction. The first one is the most stochastic component of wind speed.

1. *PGF:* Because of the roundness of the Earth and its alignment with respect to the sun, the sun heats up the Earth with uneven temperatures. Two adjacent areas with different temperatures cause a difference in pressure (pressure gradient). Pressure gradient causes air to flow from the high-pressure side to the low-pressure side to equalize the two pressures (or temperature). Wind speed increases as the PGF becomes stronger.

2. *Coriolis force:* The Coriolis force is due to the Earth's rotation. Coriolis effect is a deflection of moving air when they are viewed from a rotating reference frame such as the Earth's surface. In a reference frame with clockwise rotation (southern hemisphere), the deflection is to the left of the motion of air. For counterclockwise rotation (northern hemisphere), the deflection is to the right. Coriolis force is the strongest near the poles and zero at the equator. PGF and the Coriolis force determine the magnitude and direction of wind.

3. *Friction:* Because the surface of the Earth is rough, air friction near ground is high. Friction causes air to slow down.

Classification of wind is based on the Beaufort wind force scale. The scale was developed in 1805 by the Irish Royal Navy officer Francis Beaufort. The scale was modified several times over the years. Table 2.1 summarizes the general worldwide classification of wind. It is based on a 10-minute sustained interval. The scale 3 to 9 is the range for most modern WTs. Below 3, the wind is not strong enough to rotate the blades of the turbine. Above 9, the wind is very strong and can damage the turbine.

The strongest wind speed ever recorded was in Australia's Barrow Island on April 10, 1996. The speed of wind was 220 knots (407.44 km/h). This hurricane level wind was off Beaufort scale. Cape Farewell in Greenland is known as the "windiest region of the world."

2.1.1 Impact of Friction and Height on Wind Speed

Wind speeds decreases near ground as air friction is high. Smooth surfaces, such as water, reduce air friction. Forests or buildings slow down the wind substantially. Therefore, elevation is a key factor in determining wind speed. This is an elaborate process that

TABLE 2.1

Beaufort Scale for Wind Speed

Beaufort Scale	10-minute Sustained Winds in Knots (1 knot = 0.5144 m/s)	General Term
0	<1	Calm
1	1–3	Light air
2	4–6	Light breeze
3	7–10	Gentle breeze
4	11–16	Moderate breeze
5	17–21	Fresh breeze
6	22–27	Strong breeze
7	28–29	Moderate gale
	30–33	
8	34–40	Fresh gale
9	41–47	Strong gale
10	48–55	Whole gale
11	56–63	Storm
12	64–72	Hurricane
13	73–85	
14	86–89	
15	90–99	
16	100–106	
17	107–114	
	115–119	
	>120	

requires the knowledge of area topography as well as several meteorological parameters. An approximate method that is often used is given in Equation 2.9

$$\frac{w}{w_o} = \left(\frac{h}{h_o}\right)^{\alpha} \tag{2.9}$$

where:
α is a coefficient of friction
w is the wind speed at height h
w_o is the wind speed at known height h_o
α is function of terrain and topology of the area; typical values are $\alpha = 0.143$ for an open terrain; $\alpha = 0.4$ for a large city; and $\alpha = 0.1$ for calm water

Because power is proportional to the cube of wind speed, we can predict the power of wind as

$$\frac{P}{P_o} = \left(\frac{w}{w_o}\right)^3 = \left(\frac{h}{h_o}\right)^{3\alpha} \tag{2.10}$$

where:
P is the wind power at height h
P_o is the wind power at h_o

EXAMPLE 2.1

The wind power density at 100 m is 2.5 kW/m² when wind speed is 10 m/s. Compute the wind power density at 50 m in an open terrain.

Solution:
For an open terrain, $\alpha = 0.143$. Equation 2.10 shows the power ratios, which is the same and the power density ratio.

$$\frac{\rho}{\rho_o} = \frac{P}{P_o} = \left(\frac{h}{h_o}\right)^{3\alpha}$$

$$\rho = \rho_o \left(\frac{h}{h_o}\right)^{3\alpha} = 2.5 \times \left(\frac{50}{100}\right)^{3 \times 0.143} = 1.86 \text{ kW/m}^2$$

The National Renewable Energy Laboratory (NREL) in the United States has developed the class system, shown in Table 2.2, for wind energy sites. The data are divided into two elevations above ground level: 10 and 50 m. Sites with class 3 or greater are suitable for most utility-scale WT installations. Class 2 areas are marginal for utility-scale applications but may be suitable for rural applications.

2.1.2 Air Density

Air density is a function of air pressure, temperature, humidity, elevation, and gravitational acceleration. One of the expressions used to compute air density is

$$\delta = \frac{P_r}{\mu T} e^{(-gh/\mu T)} \tag{2.11}$$

where:
P_r is the standard atmospheric pressure at sea level (101,325 Pascal or Newton/m²)
T is the air temperature in degrees Kelvin (degree Kevin = 273.15 + degrees C)
μ is the specific gas constant, for air $\mu = 287$ Ws/(kg Kelvin)
g is the gravitational acceleration (9.8 m/s²)
h is the elevation above the sea level (m)
δ is air density in kg/m³

TABLE 2.2

Class System for Wind Energy Sites

	10 m		50 m	
Wind Power Class	**Wind Power Density (W/m²)**	**Speed (m/s)**	**Wind Power Density (W/m²)**	**Speed (m/s)**
1	0	0	0	
2	100	4.4	200	5.6
3	150	5.1	300	6.4
4	200	5.6	400	7.0
5	250	6.0	500	7.5
6	300	6.4	600	8.0
7	400	7.0	800	8.8
	1000	9.4	2000	11.9

Substituting these values into Equation 2.11 yields

$$\delta = \frac{353}{T+273} e^{[-h/29.3(T+273)]} \tag{2.12}$$

where:
  the temperature (T) in the above equation is in celsius

The equation shows that when the temperature decreases, the air is denser. Also, air is less dense at higher altitudes.

Another formula that is widely used is

$$\delta = \frac{P_r M_w 10^{-3}}{R(T+273)} \tag{2.13}$$

where:
  M_w is the molecular weight of air = 28.97 g/mol
  T is the absolute temperature (celsius)

R is the ideal gas constant = $8.2056 \cdot 10^{-5}$ $\text{m}^3 \cdot \text{atm} \cdot \text{K}^{-1} \cdot \text{mol}^{-1}$

EXAMPLE 2.2

The Green Mountain Energy Wind Farm is located in Borden and Scurry counties in Texas. The elevation of the area is 900 m above the sea level. The average wind speed of these counties is 13 m/s at 50 m above ground level. The average temperature of the area is 17°C. Compute the power density of wind at these average values.

Solution:
To compute the power density of the wind, you need to compute the air density

$$\delta = \frac{353}{17+273} e^{[-950/29.3(17+273)]} = 1.089 \, \text{kg/m}^3$$

The power density using Equation 2.8 is

$$\rho = \frac{1}{2} \delta w^3 = \frac{1}{2} 1.089 \times 13^3 = 1.196 \, \text{kW/m}^2$$

EXAMPLE 2.3

For the site in the previous example, compute the wind power passing through a sweep area of 30 m blade.

Solution:
The area of the sweep area

$$A = \pi r^2 = \pi \times 30^2 = 2827.43 \, \text{m}^2$$

The power of wind in the sweep area is

$$P_w = \rho A = 1196 \times 2827.43 = 3.381 \, \text{MW}$$

2.2 WT Blades

WT blades are somewhat similar to the general shape of airplane wings. The cross section of the wing or blade is known as "airfoil," and one of its shapes is shown at the top of Figure 2.4. The cross section of the blade has two cambers (arcs): upper and lower cambers. The upper camber is longer than the lower camber. The airfoil has a leading edge that faces the wind and trailing edge.

The airfoil has a mean camber line and a center of gravity, as shown at the middle of Figure 2.4. The points on the mean camber line are the midpoints between the upper and lower cambers. The center of gravity is located on the mean camber line. In addition, the airfoil has a cord line, which is a line connecting the leading edge to the trailing edge. All these lines are imaginary.

Air coming to the leading edge will split into two components, as shown at the bottom of Figure 2.5, with one moving along the upper camber and the other along the lower camber. The law of continuity states that the air molecules separated at the leading edge to the upper and lower camber paths meet at the trailing edge at the same time. Because the path (distance) of the upper camber is longer than that of the lower camber, the speed of air above the wing (w_1) is faster than the speed below the wing w_2.

In late 1700s, Daniel Bernoulli, a Dutch-Swiss mathematician, discovered the airlift principle known as "Bernoulli's principle." The essence of the principle is that as the velocity of air increases, pressure decreases and vice versa. Based on this, the pressure at the upper

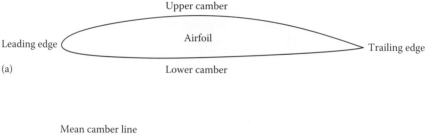

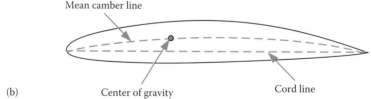

FIGURE 2.4
Airfoil: (a) shape and (b) design.

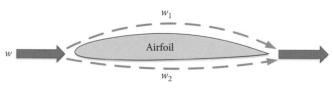

FIGURE 2.5
Airfoil and the flow of air.

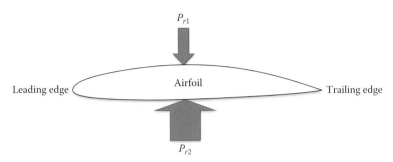

FIGURE 2.6
Bernoulli's principle.

camber P_{r1} is less than that at the lower camper P_{r2} as shown in Figure 2.6. The net pressure of the lift is

$$P_{net} = P_{r2} - P_{r1} \qquad (2.14)$$

The net pressure causes the aerodynamic force F

$$F = P_{net}A \qquad (2.15)$$

where:
  F is the aerodynamic force created by the net pressure
  P_{net} is the net pressure exerted on an object
  A is the area of the object

For the airfoil, the force at every point on the surface of the foil can be aggregated to a single component at the center of gravity of the airfoil. In our case, the aerodynamic force is called "lift force," which is perpendicular to the flow of air, as shown in Figure 2.7. For airplanes, the lift force is what makes the wing float in air. For WTs, one end of the blade (wing) is mounted on the rotating hub of the nacelle. Therefore, the lift force will create torque that causes the blade to rotate in a circular motion.

2.2.1 Angle of Attack

The aerodynamic force of the blade can be controlled by the *angle of attack*. This is α in Figure 2.8, which is the angle between the *relative wind* direction w_r and the cord line (relative wind is explained in Section 2.2.2). In the figure, the aerodynamic force F is shown for several angle of attacks. In Figure 2.8a, the airfoil has zero angle of attack, so all the

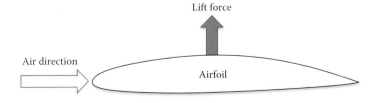

FIGURE 2.7
Lift force.

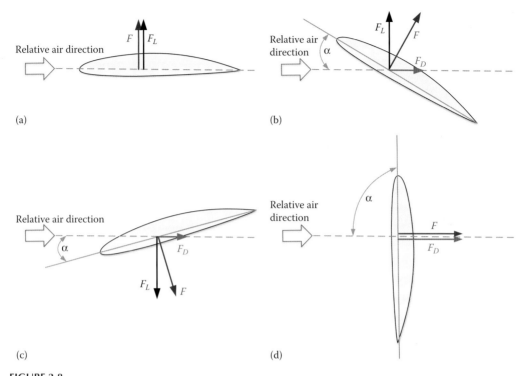

FIGURE 2.8
Aerodynamic forces and angle of attack: (a) horizontal position—all aerodynamic force is lift; (b) positive angle of attack—aerodynamic force has lift and drag; (c) negative angle of attack—lift is reversed; and (d) increasing positive angle of attack—until aerodynamic force is all drag.

aerodynamic force is lift force. In Figure 2.8b, the airfoil is at positive angle α with respect to the direction of wind in front of it. In this case, the component of the aerodynamic force is resolved into two components at the center of gravity point. One of the components is perpendicular to the direction of relative wind speed w_r in front of the blade and is called "lift force," F_L (or just lift). The component in line with the direction of relative wind speed is called "drag force," F_D (or just drag). In Figure 2.8c the angle of attack is negative. This reverses the direction of the lift force. In Figure 2.8d, the angle increases until the lift force is zero and the entire aerodynamic force is drag. During storms, the WT feathers its blades (puts them in position to minimize the lift). This is not the case in Figure 2.8d because the drag can put tremendous force on the blade causing them to deform or damage. Feathering, however, is done with a negative angle of attack. When you move the blades from positive to negative angle of attack, the lift force goes to zero before it reverses. At zero lift force, the blades are feathered.

The left force as a function of α is shown in Figure 2.9. At a negative angle α_o, the lift is zero. This is the feathering position of WTs. A larger negative angle would reverse the lift. At $\alpha = 0$, the blade provides some lift. The lift increases when the angle of attack increases up to a limit. Beyond the maximum lift level, the increase in α reduces the lift while increasing the drag force.

Keep in mind that the increase in the angle of attack increases air turbulence downstream from the WT. The lift force is substantially reduced when air is turbulent. Hence, WT located downstream from the turbine creating turbulent air captures less energy from wind.

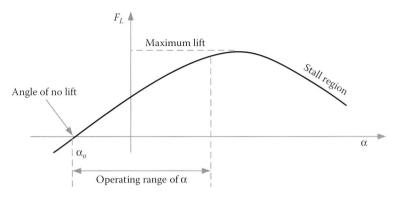

FIGURE 2.9
Lift force as a function of angle of attack.

There are two terms often used with the aerodynamic force: the lift coefficient C_L and the drag coefficient C_D. These are defined as

$$C_L = \frac{F_L}{F}$$

$$C_D = \frac{F_D}{F} \tag{2.16}$$

Both lift and drag coefficient are function of the shape of the blade and its angle of attack.

EXAMPLE 2.4

A three-blade WT is operating at a given angle of attack and the aerodynamic force exerted by wind on each blade is 2000 N. At the given angle of attack, the lift coefficient is 0.95. If the center of gravity of the blade is at 30 m from the hub, compute the torque generated by the three blades. If the blades rotate at 30 r/min, compute the mechanical power generated by the blades.

Solution:
The torque is due to the lift force that can be computed using Equation 2.16

$$F_L = C_L F = 0.95 \times 2000 = 1.9\,\text{kN}$$

The torque of one blade is

$$T = F_L r_c$$

where:
r_c is the distance from the hub to the center of gravity of the blade

For a three-blade WT, the total torque is

$$T_{total} = 3F_L r_c = 3 \times 1.9 \times 30 = 171\,\text{KNm}$$

The power generated by the blades is

$$P_{total} = T_{total}\omega_{blade} = 171\left(2\pi\frac{n_{blade}}{60}\right) = 171\left(2\pi\frac{30}{60}\right) = 537.2\,\text{kW}$$

2.2.2 Relative Wind Speed

If an instrument for measuring wind speed and wind direction is mounted on a stationary object, the readings obtained are those of the speed and direction of the true wind. If we mount the same instrument on a moving object, such as airplane, the readings will be quite different from those taken on the ground. Wind speed relative to a moving object is called "relative speed" or "apparent speed." In the top part of Figure 2.10, we have an airfoil representing an airplane moving at a speed V. If we have a true (actual) wind w in the direction shown in the figure, the wing of the airplane will see a relative wind speed w_r. You can also find the relative wind speed by the phasor sum of the head speed H to the true wind speed, where the head speed is the relative speed of a stationary object seen by the airfoil.

$$H = -V \tag{2.17}$$

$$\bar{w}_r = \bar{H} + \bar{w}$$

$$w_r = \sqrt{H^2 + w^2} \tag{2.18}$$

Now let us apply the same principle to the blade of WTs. Figure 2.11 shows the WT in front and side views. In the front view, we assumed the blades are rotating in the counter-clockwise direction at angular speed of ω. The center of gravity of each blade is assumed at a distance r_c from the center of the hub. The center of gravity point has a linear motion V that is

$$V = \omega r_c \tag{2.19}$$

In the side view, the velocity V is linear along the vertical axis passing through the center of gravity. Figure 2.12 shows the true wind velocity w, the linear velocity of the blade V and the head velocity H. As explained earlier, the relative velocity is the phasor sum of the head and true wind velocities. Note that the angle of attack α is between the cord line of the blade and the relative wind speed w_r (not the actual wind speed w). Also, note that the lift force is perpendicular to the relative wind speed w_r (not actual wind speed w) and the drag force is along the relative wind speed.

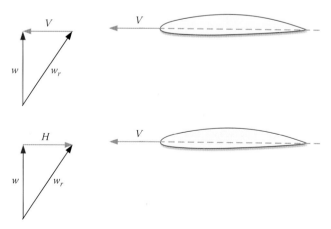

FIGURE 2.10
True and relative wind speed.

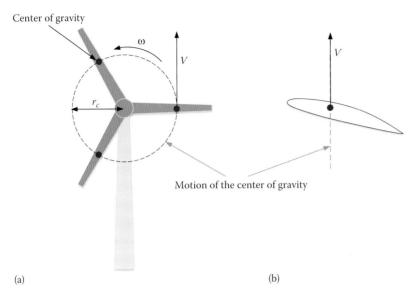

FIGURE 2.11
Motion of the center of gravity: (a) front view; (b) side view.

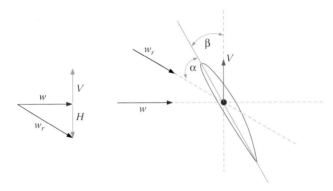

FIGURE 2.12
Relative wind speed for wind turbine blade.

EXAMPLE 2.5

The true wind speed is 15 m/s at an angle of 20° with respect to the horizontal plane. The center of gravity of the blade is 20 m from the center of the hub and is rotating at 20 r/min. Compute the relative wind speed and its direction.

Solution:
The first step is to compute the velocity of the center of gravity

$$V = \omega r_c = 2\pi \frac{20}{60} \times 20 = 41.89 \, \text{m/s}$$

The relative wind speed is

$$\bar{w}_r = \bar{H} + \bar{w} = 41.89\angle -90° + 15\angle 20° = 39.37\angle -69.02° \, \text{m/s}$$

2.2.3 Pitch Angle

Figure 2.12 shows the apparent wind speed (relative wind speed w_r) with respect to the cross section of the blade at its center of gravity. The α in the figure is the angle of attack. The angle between the cord line of the blade and the vertical line representing the linear motion of the center of gravity is called the "pitch angle," β. Note that the pitch angle is a function of the geometry of the blade and is not a function of the wind direction. The angle of attack α can be controlled by adjusting the pitch angle of the blade. For the same relative wind speed and direction, increasing the pitch angle reduces the angle of attack and vice versa. This is the main method to control the output power of WTs.

EXAMPLE 2.6

For the system in Example 2.5 if the pitch angle is 5°, compute the angle of attack.

Solution:
From Example 2.5, angle of the relative wind speed is −69.02°. Using trigonometry, the angle of attack is

$$90° - 69.02° = \alpha + \beta$$

$$\alpha = 20.98° - 5° = 15.98°$$

2.3 Coefficient of Performance

The motion of the mass of air passing through the turbine blades starts from far upstream. There is no loss of air mass by the action of the turbine blades. The mass of air can be written as

$$m = \text{vol } \delta \tag{2.20}$$

where:
  m is the mass of air
  vol is the volume of air column
  δ is air density

The volume of air is

$$\text{vol} = Awt \tag{2.21}$$

where:
  A is cross-sectional area of the air mass
  w is wind speed
  t is time period

Using Equations 2.20 and 2.21, we can write the mass of air as

$$m = Aw\delta t \tag{2.22}$$

Because the mass is constant, the mass flow rate f_{mass} is also constant for the same period of time.

$$f_{\text{mass}} = \frac{m}{t} = Aw\delta \tag{2.23}$$

Because f_{mass} is constant, the area times wind speed is constant at any distance from the blade. Let w_u be the wind speed at the far upstream from the WT, which is the steady-state wind speed in far front of the turbine. Also, assume that the wind speed at the far downstream is w_d and at the blade is w. Hence, we can conclude from Equation 2.23 that

$$A_u w_u = A_d w_d = A_{\text{blade}} w \tag{2.24}$$

where:
  A_u is the cross-sectional area of the air mass at far upstream distance
  A_d is the cross-sectional area of the air mass at far downstream distance
  A_{blade} is the cross-sectional area of the air mass through the turbine blades (the same as the sweep area of the blade)

But what is causing the change in wind speed? When wind mass passes through the blades, part of its kinetic energy is harvested by the blades. Hence, the kinetic energy of the air mass behind the turbine is less than that in front of it. The energy harvested by the blade is the difference between the upstream kinetic energy of wind KE_u and the downstream kinetic energy of wind KE_d (assuming no energy losses due obstacles or air particles).

$$KE_{\text{blade}} = KE_u - KE_d \tag{2.25}$$

where:

$$KE_u > KE_d \tag{2.26}$$

We can write Equation 2.26 in terms of mass and speed

$$\frac{1}{2} m w_u^2 > \frac{1}{2} m w_d^2 \tag{2.27}$$

Because the mass of air is constant, the upstream wind speed must be larger than the downstream wind speed.

$$w_u > w_d \tag{2.28}$$

This is why the presence of WT in a large moving air mass modifies the local air speed. Equation 2.24 is represented in Figure 2.13. The top graph in the figure shows the envelope of the air mass passing through a WT. Three locations are considered. At the far upstream, the wind speed w_u is the highest and the cross section of the air mass envelope A_u is the smallest. The downstream wind speed w_d is the lowest, so the cross section at this location A_d is the largest. At the turbine, the wind speed is in between w_u and w_d. Therefore, the air mass cross section is also between A_u and A_d. The bottom part of the figure shows the wind speed at the various distances.

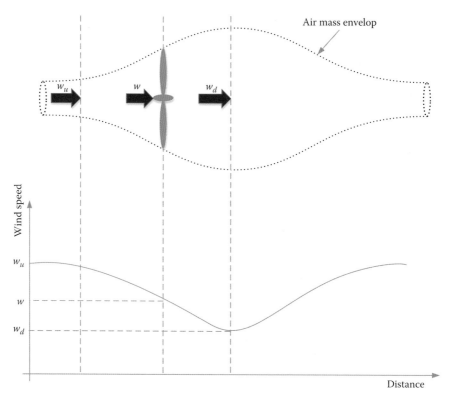

FIGURE 2.13
Circular tube of air flowing through a wind turbine.

According to the conservation of linear momentum theory and Bernoulli's principal, the speed of wind at the blade is the average of w_u and w_d.

$$w = \frac{w_u + w_d}{2} \tag{2.29}$$

Based on Equation 2.29, we can compute the mass of air

$$m = \delta A_{\text{blade}} w t = \delta A_{\text{blade}} \left(\frac{w_u + w_d}{2} \right) t \tag{2.30}$$

Substituting Equation 2.30 into 2.25 yields

$$KE_{\text{blade}} = \frac{1}{2} m w_u^2 - \frac{1}{2} m w_d^2 = \frac{1}{2} \delta A_{\text{blade}} \left(\frac{w_u + w_d}{2} \right) \left(w_u^2 - w_d^2 \right) t \tag{2.31}$$

and the power captured by the blade is the energy divided by time

$$P_{\text{blade}} = \frac{1}{2} \delta A_{\text{blade}} \left(\frac{w_u + w_d}{2} \right) \left(w_u^2 - w_d^2 \right) \tag{2.32}$$

Defining the ratio of the downstream to the far upstream wind speed as

$$\gamma = \frac{w_d}{w_u} \qquad (2.33)$$

Then, we can rewrite Equation 2.32 as

$$P_{\text{blade}} = \frac{1}{2}\delta A_{\text{blade}}\left(\frac{w_u + \gamma w_u}{2}\right)(w_u^2 - \gamma^2 w_u^2) = \frac{1}{4}\delta A_{\text{blade}}w_u^3(1+\gamma)(1-\gamma^2) \qquad (2.34)$$

Defining

$$P_w = \frac{1}{2}\delta A_{\text{blade}}w_u^3 \qquad (2.35)$$

P_w is called the "wind power," which is the power computed using the sweeping area of the blade and the far upstream wind speed. Keep in mind that P_w is not equal to the far upstream power P_u, the downstream power P_d, or the blade power P_{blade}. Because

$$P_u = \frac{1}{2}\delta A_u w_u^3 \qquad (2.36)$$

then

$$P_w = P_u \frac{A_{\text{blade}}}{A_u} \qquad (2.37)$$

Similarly, because

$$P_d = \frac{1}{2}\delta A_d w_d^3 \qquad (2.38)$$

Hence,

$$P_w = P_d \frac{A_{\text{blade}}}{A_d}\left(\frac{w_u}{w_d}\right)^3 \qquad (2.39)$$

When a site is evaluated for potential WT installation, wind power density is used as a measure for the expected production. Wind power density ρ is defined as the wind power P_w per unit of a sweep area.

$$\rho = \frac{P_w}{A_{\text{blade}}} \qquad (2.40)$$

Normally ρ between 300 and 500 indicates a good wind site.

Equation 2.34 can be written as

$$P_{\text{blade}} = P_w C_p$$

$$C_p = \frac{P_{\text{blade}}}{P_w} \qquad (2.41)$$

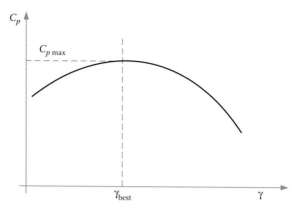

FIGURE 2.14
Coefficient of performance as a function of wind-speed ratio.

where:

$$C_p = \frac{1}{2}(1+\gamma)(1-\gamma^2) \tag{2.42}$$

C_p is known as the "coefficient of performance." It represents the amount of wind power that is captured by the blades. To compute the maximum value of C_p, we need to equate the derivative of Equation 2.42 to zero.

$$\frac{\partial C_p}{\partial \gamma} = \frac{1}{2}(1-2\gamma-3\gamma^2) = 0 \tag{2.43}$$

Solving Equation 2.43 leads to the best value of γ

$$\gamma_{best} = \frac{1}{3} \tag{2.44}$$

Substituting Equation 2.44 into 2.42 leads to the maximum C_p

$$C_{p\,max} = \frac{1}{2}(1+\gamma_{best})(1-\gamma_{best}^2) \approx 0.593 \tag{2.45}$$

This value of the maximum C_p is known as the "Betz's limit." It is a theoretical maximum developed by Albert Betz in 1920. It shows that we can capture up to 59.3% of the wind power if the downwind speed is one-third of the far upwind speed. There is no WT today that can achieve this value.

Equation 2.42 is plotted in Figure 2.14. Note that the coefficient of performance is reduced when we move away from γ_{best}.

EXAMPLE 2.7

A WT has a mass flow rate of 20,000 kg/s. The upwind speed is 20 m/s and the downwind speed is 18.7 m/s. Compute the following:

1. Diameter of the air mass boundary in the upwind and downwind regions
2. Power in the upwind and downwind areas
3. Power captured by the blades

4. Coefficient of performance computed using the upwind and downwind powers
5. Coefficient of performance using wind speeds

Solution:

1. Equation 2.22 can be used to compute the cross-sectional area of the air mass passing through the turbine

$$m = Aw\delta t$$

where the mass flow rate is

$$f_{mass} = \frac{m}{t} = Aw\delta$$

Assuming the air density to be 1 kg/m³, the cross-sectional areas of air mass at the far upstream and far downstream are

$$A_u = \frac{f_{mass}}{w_u \delta} = \frac{20,000}{20} = 1000 \, \text{m}^2$$

$$A_d = \frac{f_{mass}}{w_d \delta} = \frac{20,000}{18.7} = 1070 \, \text{m}^2$$

$$A_{blade} = \frac{f_{mass}}{w\delta} = \frac{f_{mass}}{\left[(w_u + w_d)/2\right]\delta} = \frac{20,000}{19.35} = 1033.6 \, \text{m}^2$$

The diameters of the upwind and downwind regions are as follows:

$$d_u = 2 \times \sqrt{\frac{1000}{\pi}} = 35.7 \, \text{m}$$

$$d_d = 2 \times \sqrt{\frac{1070}{\pi}} = 37 \, \text{m}$$

2. Using Equations 2.36 and 2.38

$$P_u = \frac{1}{2}\delta A_u w_u^3 = \frac{1}{2} \times 1000 \times 20^3 = 4 \, \text{MW}$$

$$P_d = \frac{1}{2}\delta A_d w_d^3 = \frac{1}{2} \times 1070 \times 18.7^3 = 3.5 \, \text{MW}$$

3. The power captured by the blade is

$$P_{blade} = P_u - P_d = 4 - 3.5 = 0.5 \, \text{MW}$$

4. The coefficient of performance using Equation 2.41 is

$$C_p = \frac{P_{\text{blade}}}{P_w} = \frac{P_{\text{blade}}}{P_u} \frac{A_u}{A_{\text{blade}}} = \frac{0.5}{4} \frac{1000}{1033.6} = 0.121$$

5. The coefficient of performance computed by wind speeds using Equation 2.42 is

$$C_p = \frac{1}{2}(1+\gamma)(1-\gamma^2)$$

where:

$$\gamma = \frac{w_d}{w_u} = \frac{18.7}{20} = 0.935$$

Hence,

$$C_p = \frac{1}{2}(1+0.935)(1-0.935^2) = 0.121$$

2.3.1 Tip-Speed Ratio

Figure 2.15 shows a frontal view of the rotating blade. The linear velocity at the tip of the blade is known as the "tip velocity," V_{tip}, which is a function of the angular speed of the blade ω and the length of the blade r.

$$V_{\text{tip}} = \omega r = 2\pi n r \tag{2.46}$$

where:
n is the number of revolutions the blade makes in one second

$$n = \frac{V_{\text{tip}}}{2\pi r} \tag{2.47}$$

The WT is often designed to have its tip velocity faster than wind speed to allow the turbine to generate electricity even at low wind speeds. However, a high tip speed produces audible noise. In most turbine design, the tip speed is limited to about 80 m/s in areas with noise restrictions. The ratio of the tip velocity V_{tip} to the wind speed w_u is known as the "tip-speed ratio" (TSR) (λ).

$$TSR = \lambda = \frac{V_{\text{tip}}}{w} \tag{2.48}$$

The TSR is an easier measure for the coefficient of performance than γ because the TSR requires just the measurement of wind speed at the turbine. No need for wind speed measurements far away from the turbine. Modern WTs have TSRs range of about 5–10.

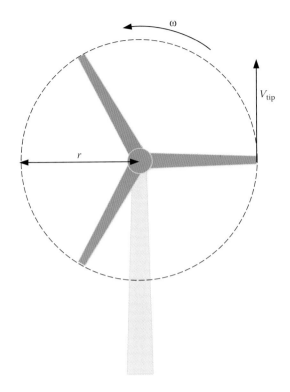

FIGURE 2.15
Tip velocity.

2.3.2 Blade Power

If we compute the blade power using the exerted torque on the blade, we get

$$P_{\text{blade}} = \omega T_{\text{blade}} = \omega F_L r_c \tag{2.49}$$

where:
T_{blade} is the torque exerted on the blades
F_L is the equivalent lift force computed at the center of gravity of the blade
r_c is the distance between the center of gravity and the hub

The power in the upstream wind is

$$P_u = F_u w_u \tag{2.50}$$

where:
F_u is the force of upstream wind
w_u is the upstream wind speed

Hence, the ratio of the two powers is

$$\frac{P_{\text{blade}}}{P_u} = \frac{\omega F_L r_c}{F_u w_u} \tag{2.51}$$

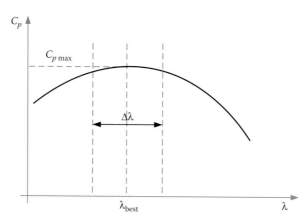

FIGURE 2.16
Coefficient of performance as a function of the tip-speed ratio.

P_u can be substituted by P_w in Equation 2.37

$$\frac{P_{\text{blade}}}{P_w} = \frac{\omega F_L r_c}{F_u w_u} \frac{A_u}{A_{\text{blade}}} = \frac{V_{\text{tip}} F_L r_c}{F_u w_u r} \frac{A_u}{A_{\text{blade}}} = \lambda \frac{F_L r_c}{F_u r} \frac{A_u}{A_{\text{blade}}} \tag{2.52}$$

Equation 2.52 represents the coefficient of performance, hence

$$C_p = \frac{P_{\text{blade}}}{P_w} = \lambda \frac{F_L r_c}{F_u r} \frac{A_u}{A_{\text{blade}}} \tag{2.53}$$

The equation shows that the coefficient of performance is a function of λ as well as the various forces and areas within the air mass envelope. The lift force is also a function of the pitch angle as well as the TSR. All these variables change with wind speed. The coefficient of performance is a nonlinear function of λ and has the general shape in Figure 2.16.

A well-designed system operates the WT at or near the maximum C_p as shown by $\Delta\lambda$ in Figure 2.16. The power captured is given in Equation 2.41, which can be written as a function of wind speed at the turbine blade

$$P_{\text{blade}} = \frac{1}{2}\delta A_{\text{blade}} w_u^3 C_p = \frac{1}{2}\delta \frac{A_{\text{blade}}^4}{A_u^3} w^3 C_p \tag{2.54}$$

Figure 2.17 shows the blade power as a function of wind speed. The operating range is divided into four regions. The first region is for wind speed below the cut-in speed w_{min}. The second region is for wind speed between w_{min} and the speed w_B at which the turbine reaches its rated power production. The third region is for wind speed higher than w_B, while the power of the turbine is controlled to stay at its rated value. The fourth region is for wind speed higher than the design limit (cut-out speed w_{max}).

At point A, the wind speed is high enough to start generating electricity. Between A and B, the turbine output is a function of the cube of wind speed as well as the pitch angle. During this region, the pitch angle is adjusted to operate the turbine at its maximum C_p to harvest as much energy from wind as possible. The power at point B is the rated output power of the turbine. If wind speed exceeds w_B, the blades are adjusted to spell some of the wind energy to operate the turbine at its rated power.

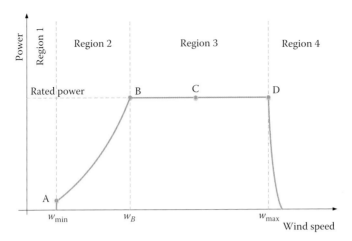

FIGURE 2.17
Output power of wind turbine.

The blade and hub experience tremendous stress due to centripetal force imposed on them by the wind. The centripetal force from the spinning blade increases proportional to the square of the rotation speed. This makes the structure sensitive to over speeding. Therefore, when the speed of wind reaches the maximum design limit of the turbine (also known as the "cut-out speed" or "maximum speed," w_{max}), the turbine is aerodynamically stalled (known as "feathering") to make C_p near zero and the mechanical brakes are applied to stop the rotation of the blades. This is point D in Figure 2.17.

As seen in Equation 2.53, the coefficient of performance is a function of the lift force. The lift force, in turn, is a function of the pitch angle β. Hence, C_p is a function of TSR λ and lift force (or pitch angle).

$$C_p = f(F_L, \lambda) = f(\beta, \lambda) \tag{2.55}$$

Figure 2.18 shows a family of C_p curves for several pitch angles

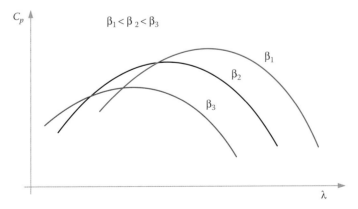

FIGURE 2.18
Coefficient of performance as a function of pitch angle.

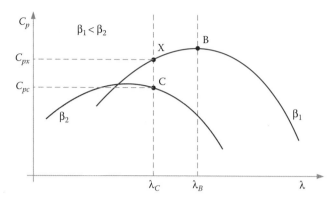

FIGURE 2.19
Coefficient of performance as a function of pitch angle.

Now let us assume that we have reached point B in Figure 2.17 at wind speed w_B, which corresponds to point B in Figure 2.19. Assume that the wind speed increases, which reduces the *TSR* from λ_B to λ_C. With no pitch angle control, the coefficient of performance is reduced to C_{pX} in Figure 2.19. If the blade power at point X is still at the rated value, there is no need to do anything. However, if the blade power exceeds the rated value, the coefficient of performance must be reduced. This can be done by increasing the pitch angle (reducing the angle of attack and reducing the lift force) to maintain the power at the rated value. This is point C in Figure 2.19, which corresponds to a point between B and D in Figure 2.17.

The coefficient of performance can be empirically computed for a given WT design. One of the equations widely used is

$$C_P = k_1(\Lambda - k_2\beta - k_3\beta^3 - k_4)e^{-\Lambda k_5} \tag{2.56}$$

$$\Lambda = \frac{1}{\lambda + k_6\beta} - \frac{k_7}{1 + \beta^3} \tag{2.57}$$

where:
k_1 to k_7 are constants unique to any given turbine design

EXAMPLE 2.8

A turbine has an upstream wind speed of 10 m/s. The TSR of the turbine is 5. The length of the turbine blade is 50 m and the pitch angle is 10°. The constants of the coefficient of performance are as follows:

$$k_1 = 20$$

$$k_2 = 0.1$$

$$k_3 = 0.002$$

$$k_4 = 0.003$$

$$k_5 = 15$$

$$k_6 = 1$$

$$k_7 = 0.02$$

Compute the coefficient of performance and the power captured by the blades.

Solution:
Equations 2.56 and 2.57 can be used to compute the coefficient of performance

$$\Lambda = \frac{1}{\lambda + k_6\beta} - \frac{k_7}{1+\beta^3} = \frac{1}{5+(\pi/18)} - \frac{0.02}{1+(\pi/18)^3} = 0.1734$$

$$C_P = 20\left[0.1734 - 0.1\left(\frac{\pi}{18}\right) - 0.002\left(\frac{\pi}{18}\right)^3 - 0.003\right]e^{-0.1734\times 15} = 0.227$$

The power can be computed using Equations 2.35 and 2.41

$$P_{\text{blade}} = P_w C_p = \frac{1}{2}\delta A_{\text{blade}} w_u^3 C_p = \frac{1}{2}\left(\pi\times 50^2\right)10^3 \times 0.227 = 891.4\,\text{kW}$$

2.4 Separation of WTs

Wind farms are located all over the world. Clustering WTs in one location (wind farms) makes engineering and economic sense because of several reasons:

- Reduces the installation costs as the expensive heavy equipment do not have to move over long distances
- Reduces operation costs as system operators are placed in one location
- Reduces maintenance costs
- Simplifies grid connection

The drawback of the clusters, besides the high impact of wind variability, are as follows:

- Wind slows down as it passes through the blades. Thus, available wind power to downwind machines is reduced.
- Wing passing through blades creates turbulences. Close turbines located downstream from other turbines cannot efficiently capture energy from the turbulent wind.

Because of the above reasons, WTs must be adequately separated to allow the wind turbulences to damp out and wind speed recovers before it reaches the next turbine. A sample of WTs cluster is shown in Figure 2.20.

FIGURE 2.20
Wind turbines array. (Courtesy of the US National Renewable Energy Laboratory.)

One arrangement of WTs is the square configuration shown in Figure 2.21. D is the distance between two adjacent towers. The separation between the two turbines is the minimum distance between the tips of the two adjacent turbine blades. The separation factor S is defined as

$$S = \frac{D}{2r} \qquad (2.58)$$

where:
  r is the length of the blade

The separation plays a key role in the amount of power that can be captured by the turbines. Figure 2.22 shows the array efficiency versus separation. The array efficiency η_{array} is a measure of how much of the wind energy is available to the turbines in the array. As seen in the figure, the 2×2 array requires less separation for the same efficiency.

The array efficiency can be computed empirically by a curve fitting formula such as

$$\eta_{array} = 100(1 - ae^{-bS}) \qquad (2.59)$$

where:
  a and b are constants values that depend on the number of turbines in the array

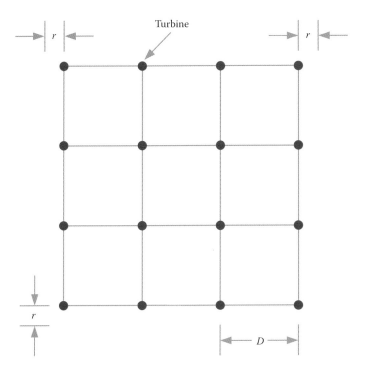

FIGURE 2.21
Square arrangement of a turbine array.

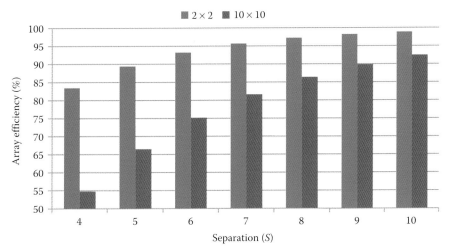

FIGURE 2.22
Wind array power versus separation.

The acquired land depends on the number of turbines, the length of the blade, and the separation factor. Using Figure 2.21, we can calculate the minimum land use as

$$A_{land} = [(x-1)D + 2r]^2 \qquad (2.60)$$

where:
   x is the number of turbine in one row

EXAMPLE 2.9

A wind developer acquires a 10 × 10 km land to install WTs of 50 m blade length. To achieve a separation of 8, how many WTs can be installed at the site?

Solution:
Equation 2.58 can be used to compute the distance between towers

$$D = 2rS = 100 \times 8 = 800\,m$$

The number of WTs in each row or column can be computed by Equation 2.60

$$A_{land} = [(x-1)D + 2r]^2$$

$$10^8 = [(x-1) \times 800 + 2 \times 50]^2$$

$$x = 1 + int\left(\frac{9900}{800}\right) = 13\,turbines$$

For the whole site, the number of turbines is

$$n_{total} = 13 \times 13 = 169\,turbines$$

EXAMPLE 2.10

For the wind farm in the previous example, compute the power production per land area when the wind power density at the hub is 400 W/m², the coefficient of performance is 0.3, and the overall efficiency of the turbine-generator system is 85%. Assume the array efficiency is 74%.

Solution:
Equation 2.40 can be used to compute the wind power

$$P_w = \rho A_{blade} = 400(\pi \times 50^2) = 3.14\,MW$$

The output power of the turbine is

$$P_{\text{out}} = P_{\text{blade}}\eta = P_w C_p \eta = 3.14 \times 0.3 \times 0.85 = 800\,\text{kW}$$

The total power of the turbines in the farm is

$$P_{\text{total}} = P_{\text{out}}\,n_{\text{total}} = 800 \times 169 = 135.2\,\text{MW}$$

Because of the array efficiency, the output power of the farm is

$$P_{\text{farm}} = P_{\text{out}}\,\eta_{\text{array}} = 135.2 \times 0.74 = 100.05\,\text{MW}$$

The power production per land area

$$\frac{P_{\text{farm}}}{A_{\text{land}}} = \frac{100.05}{10 \times 10} \approx 1.0\,\text{MW/km}^2$$

The square configuration is suitable for wind farms exposed to variable wind directions. When the prevailing wind direction is consistent (such as near shores or mountain passes), the diagonal arrangement, shown in Figure 2.23, is often used. The common ranges of D_1 and D_2 are as follows:

$$6r \leq D_1 \leq 10r$$

$$10r \leq D_2 \leq 20r \tag{2.61}$$

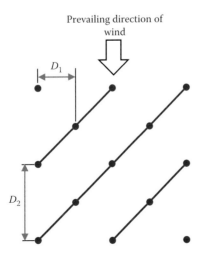

FIGURE 2.23
Diagonal arrangement of turbines array.

Exercise

1. A three-blade wind turbine captures 1 MW from wind moving horizontally with respect to the plane. If the blades rotate at 20 r/min, compute the torque exerted by each blade.

2. A three-blade wind turbine captures 1 MW from wind moving horizontally with respect to the plane. If the upwind speed is 15 m/s and the coefficient of performance is 10%, compute the length of the blade.

3. A wind turbine has 50 m tower. The turbine is installed offshore in calm water where the wind power at 10 m height is 450 kW. Compute the wind power at the hub level.

4. The true wind speed is 15 m/s at zero angle with respect to the horizontal plane. A wind turbine blade with a center of gravity of 30 m from the center of the hub is rotating at 20 r/min. Compute the relative wind speed.

5. If wind speed at the blade of a turbine is 20 m/s and the downwind speed is 15 m/s, compute the upwind speed.

6. A wind turbine with 10 m blade length has upwind speed of 20 m/s and downwind speed of 10 m/s. Compute the power that is captured by the blade.

7. A wind turbine has a pitch angle of 5°. The upwind speed is 15 m/s moving horizontally with respect to the plane. The center of gravity of the blade is 20 m from the center of the hub. The blades are rotating at 30 r/min. Compute the angle of attack.

8. The true wind speed is 15 m/s at an angle of 20° with respect to the horizontal plane. A wind turbine blade with a center of gravity of 20 m from the center of the hub is rotating at 20 r/min. Compute the relative wind speed.

9. The relative wind speed is 15 m/s at an angle of –20° with respect to the horizontal plane. If the pitch angle of the wind turbine is 30°, compute the angle of attack.

10. If the wind speed at the blade of a turbine is 2 m/s and the downwind speed is 15 m/s, compute the approximate upwind speed.

11. A wind turbine with 10 m blade length has upwind speed of 20 m/s and downwind speed of 10 m/s. Compute the power that can be captured by the blade.

12. The coefficient of performance of a wind turbine is 20% at a given pitch angle when the upwind speed is 10 m/s. The length of the blade is 50 m. Compute the flow rate of the air mass and the power captured by the blades.

13. A wind turbine has a mass flow rate of 10^5 kg/s. The upwind speed is 10 m/s and the downwind speed is 8 m/s. Compute the following:

    a. Power in the upwind and downwind areas

    b. The power captured by the blades

    c. The coefficient of performance computed using the upwind and downwind powers

    d. The coefficient of performance using the wind speeds

14. At a wind speed of 10 m/s the tip-speed ratio of a wind turbine is 5. The length of the turbine blade is 50 m and the pitch angle is 10°. The constants of the coefficient of performance are as follows:

$$k_1 = 20$$

$$k_2 = 0.1$$

$$k_3 = 0.002$$

$$k_4 = 0.003$$

$$k_5 = 15$$

$$k_6 = 1$$

$$k_7 = 0.02$$

Plot the coefficient of performance for pitch angle equal to 0°, 10°, 20°, and 30° for the tip-speed ratio up to 20.

15. A wind developer acquires a 10×10 km land to install wind turbines of 50 m blade length. To achieve a separation of 8, how many wind turbines can be installed at the site?

16. For the wind farm in the previous example, compute the power production per land area. Assume that the wind power density at the hub is 400 W/m², the coefficient of performance is 0.3, and the overall efficiency of the turbine-generator system is 85%.

17. What are the main variables determining the amount of energy captured by wind turbines?

18. What are the factors determining wind speed?

19. What is the pressure gradient force?

20. What is the Coriolis force?

21. What is the approximate range of wind speed that is suitable for wind turbines?

22. How does wind speed change with height and friction?

23. What are the factors determining air density?

24. What is the cord line?

25. What is the mean camber line?

26. How is left occur on an airfoil?

27. What is the difference between angle of attack and pitch angle?

28. What is the left coefficient?

29. What is the drag coefficient?

30. What is the difference between wind speed and relative wind speed?

31. What is the head speed?

32. How does pitch angle change the power captured by the blade?

33. What is the coefficient of performance?

34. What is the tip-speed ratio?

35. What are the variables that determine the coefficient of performance?

36. What is the theoretical maximum for the coefficient of performance?

37. Under what conditions do a turbine spills wind?

38. What are the factors determining the separation of wind turbines?

3

Wind Statistics

Before wind turbines are installed at a given site, measurements of wind speed over a period of time are made. Once the measurements are obtained and analyzed, the potential for energy production can be evaluated. Based on statistical methods wind is highly variable. Normally, the first step is to measure the frequency of wind speeds at the site, as shown in Figure 3.1. The horizontal axis is for wind speeds and the vertical axis is for the number of hours each wind speed value occurs in one year. In statistical terms, the data on the vertical axis is known as the "frequency of occurrence," n. The wind speed on the horizontal axis can be represented by an index. For example, the index for wind speed = 0 is 1, the index for wind speed = 2.0 m/s is 2, and so on. The frequency of wind speed of 6.0 m/s as shown in Figure 3.1 is 1100 h. This can be written as

$$n_4 = 1100 \, \text{h} \tag{3.1}$$

In one year, the sum of all frequencies must equal to the number of hours in one year

$$\sum_{i=1}^{\infty} n_i = 8765 \, \text{h} \tag{3.2}$$

Instead of using the raw frequency, we can use a relative term known as the "probability," p, which is a normalized frequency. The probability of any wind speed is its frequency divided by the total number of hours in a year (8765 h).

$$p_i = \frac{n_i}{8765} \tag{3.3}$$

The sum of all probabilities for all wind speeds must equal to 1.

$$P = \sum_{i=1}^{\infty} p_i = \sum_{i=1}^{\infty} \frac{n_i}{8765} = 1 \tag{3.4}$$

The frequency distribution in Figure 3.1 can now be replaced by the probability distribution in Figure 3.2.

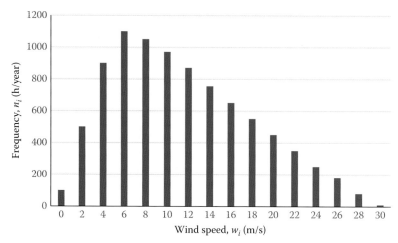

FIGURE 3.1
Frequency distribution of wind speed at a given site.

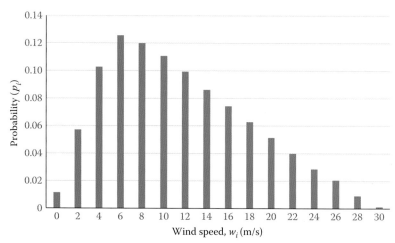

FIGURE 3.2
Probability distribution of wind speed at a given site.

3.1 Average Variance and Standard Deviation

There are three important variables that provide substantial information on wind speed: average, variance, and standard deviation. The average wind speed (w_{ave}) is just the mean value of all wind speeds. It is also known as the "expected value." Therefore, the average wind speed at a site is

$$\mathrm{E}[w] = w_{ave} = \frac{1}{N} \sum_{i=1}^{\infty} n_i\, w_i$$

$$\mathrm{E}[w] = w_{ave} = \sum_{i=1}^{\infty} p_i\, w_i$$

(3.5)

where:

E[w] is the expected value of wind speed

N is the hours in the period of data. If the data is collected over one year, $N = 8765$ h

The dispersion of wind speeds from its average is known as the "variance," Var; it is defined by

$$Var = \frac{1}{N} \sum_{i=1}^{\infty} n_i \, (w_i - w_{ave})^2 \tag{3.6}$$

The reason for squaring the difference in Equation 3.6 is to not allow the positive and negative deviations from canceling each other. To somehow account for the squaring, the standard deviation σ is used

$$\sigma = \sqrt{Var} = \sqrt{\frac{1}{N} \sum_{i=1}^{\infty} n_i \, (w_i - w_{ave})^2} \tag{3.7}$$

A small σ indicates that most wind speeds are close to the average (excellent if the average wind speed is suitable for energy production). However, large σ indicates that wind speeds are spread wide over a large range.

EXAMPLE 3.1

The following measurements were made at a site at an elevation of 100 m.

Range	Average of Range, w_i (m/s)	Frequency, n_i (h)
0–1.0	0.5	1000
1.0–2.0	1.5	1300
2.0–3.0	2.5	1100
3.0–4.0	3.5	1800
4.0–5.0	4.5	1700
5.0–6.0	5.5	1200
6.0–7.0	6.5	600
7.0–10.0	8.5	65

Compute the average wind speed and the standard deviation.

Solution:

Equation 3.5 can be used to compute the average wind speed.

$$w_{ave} = \frac{1}{N} \sum_{i=1}^{\infty} n_i \, w_i = \frac{1}{8765}(1000 \times 0.5 + 1300 \times 1.5 + \cdots + 65 \times 8.5) = 3.45 \text{ m/s}$$

The variance is

$$Var = \frac{1}{N} \sum_{i=1}^{\infty} n_i (w_i - w_{ave})^2$$

$$= \frac{1}{8765}\left[1000 \times (3.45 - 0.5)^2 + 1300 \times (3.45 - 1.5)^2 + \cdots + 65 \times (3.45 - 8.5)^2 \right]$$

$$= 3.285 \text{ m}^2/\text{s}^2$$

The standard deviation is

$$\sigma = \sqrt{\text{Var}} = \sqrt{3.285} = 1.81\,\text{m/s}$$

3.2 Cumulative Distribution Function

As given in Equation 3.4, the sum of all probabilities for all wind speeds is 1. This is known as "cumulative probability" of all wind speeds. However, wind turbines cannot operate at all wind speeds; they are designed to operate when wind speed is within a specific range (from x to y), as shown in Figure 3.3. The cumulative probability for this range is

$$F_{xy} = p(w_x \le w_i \le w_y) = \sum_{i=x}^{i=y} p_i \tag{3.8}$$

The accumulate probability F_{xy} is known as the "cumulative distribution function" (CDF).

EXAMPLE 3.2

For the data in Example 3.1, the wind turbines can generate electricity when wind speed range is 6–8 m/s. What is the probability of generating electricity in one year?

Solution:
Equation 3.8 can be used to compute the CDF

$$F_{6-8} = p(w_6 \le w_i \le w_8) = \sum_{6}^{8} p_i = \frac{1}{8765}(1200 + 600 + 65) = 21.28\%$$

The site can produce electricity 21.28% of the time annually.

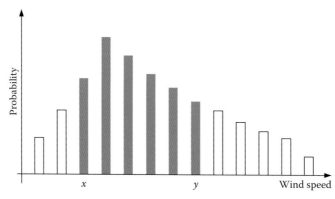

FIGURE 3.3
Probability distribution of wind speeds that generate electricity.

3.3 Probability Density Function

Using raw data for analysis is not simple because the measurements can be immense in size, difficult to manage, and difficult to extract specific information from it. In addition, data may be missing for certain ranges. To address these problems, mathematical functions representing the characteristic of the data are used instead of the row data. These functions are known as the "probability density functions" (PDF). An example of a PDF is shown in Figure 3.4.

If the measurements fit a PDF, there is no need to keep or maintain the raw data. The function can be used in future analyses instead of the raw data. In addition, with accurate PDF, we can interpolate the data when certain ranges of wind speeds are missing.

The PDF function for wind speed (also denoted $f[w]$) can also provide accurate statistical properties of the raw data. For example, the cumulative value of $f(w)$ over all wind speeds is equal to 1.

$$\int_{w=0}^{\infty} f(w)\,dw = 1 \tag{3.9}$$

The CDF for a range from x to y is

$$F_{xy} = \int_{w_x}^{w_y} f(w)\,dw \tag{3.10}$$

The average wind speed is

$$E[w] = w_{ave} = \int_{w=0}^{\infty} w\,f(w)\,dw \tag{3.11}$$

The variance is

$$Var = \int_{w=0}^{\infty} (w - w_{ave})^2\,f(w)\,dw \tag{3.12}$$

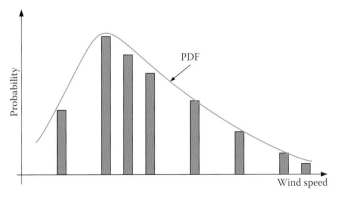

FIGURE 3.4
Probability density function.

The standard deviation is

$$\sigma = \sqrt{\int_{w=0}^{\infty} (w - w_{ave})^2 \, f(w) \, dw} \tag{3.13}$$

The major question is how to find a suitable PDF function? This is normally done by curve fitting the data to several known-shape functions. Because wind speeds do not follow normal distribution characteristics, the PDF function should peak at wind speed less than w_{ave}, as shown in Figure 3.4. Researchers have found several suitable functions; two of them are commonly used: Weibull distribution function and Rayleigh distribution function.

3.3.1 Weibull Distribution Function

The Weibull distribution function has the form

$$f(w) = \frac{k}{c^k} w^{k-1} e^{-(w/c)^k} \tag{3.14}$$

where:
   c is called the "scale parameter" that can mainly adjust the magnitude of the function
   k is called the "shape parameter" that can mainly shift the peak of the function

The effects of these two parameters are shown in Figure 3.5. Their values are normally obtained by curve fitting the raw data to the function in Equation 3.14. Instead, an approximate method can be used where the two parameters are obtained using the values of the average and standard deviation of the raw data.

$$k \approx \left(\frac{\sigma}{w_{ave}} \right)^{-1.086} ; \quad \text{for } 1 \le k \le 10 \tag{3.15}$$

$$c \approx \frac{w_{ave}}{\Gamma[1 + (1/k)]} \tag{3.16}$$

where $\Gamma(y)$ is the Gama function of y defined as

$$\Gamma(y) = \int_{x=0}^{\infty} x^{y-1} e^{-x} dx \tag{3.17}$$

Assume that

$$y = 1 + \frac{1}{k} \tag{3.18}$$

then

$$\Gamma\left(1 + \frac{1}{k}\right) = \int_{x=0}^{\infty} x^{1/k} e^{-x} dx \tag{3.19}$$

Most scientific calculators with statistics features can provide the value of the Gama function in Equation 3.19.

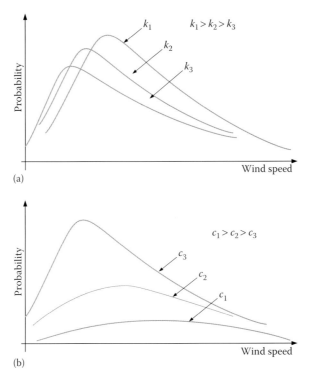

FIGURE 3.5
Effect of Weibull distribution function parameters: (a) shape parameter and (b) scale parameter.

With Weibull function, the CDF, variance, and standard deviation are as follows:

$$F_{xy} = p(w_x \leq w_i \leq w_y) = \int_{w_x}^{w_y} f(w)dw = e^{-(w_x/c)^k} - e^{-(w_y/c)^k} \tag{3.20}$$

$$\text{Var} = c^2 \left[\Gamma\left(1+\frac{2}{k}\right) - \Gamma\left(1+\frac{1}{k}\right) \right] \tag{3.21}$$

$$\sigma = c \sqrt{\left[\Gamma\left(1+\frac{2}{k}\right) - \Gamma\left(1+\frac{1}{k}\right) \right]} \tag{3.22}$$

EXAMPLE 3.3

A Wiebull function representing the wind speed at a given site has $c = 5$ and $k = 1.2$. Compute the number of hours per year when wind speed is 4 m/s or greater.

Solution:
Use Equation 3.20 to compute the CDF assuming $w_y = \infty$

$$F_{xy} = \int_{4}^{\infty} f(w)dw = e^{-(4/5)^k} = 0.485$$

The availability of wind speed of 4 m/s or higher is $= 0.485 \times 8765 = 4251$ h/year.

3.3.2 Rayleigh Distribution Function

Weibull distribution function is a two-parameter equation. A simpler function, but probably less accurate, is the Rayleigh function (also known as "chi-2 function"), which is described by one parameter only, the average wind speed. The shape of this PDF function is

$$f(w) = \frac{\pi}{2} \frac{w}{w_{\text{ave}}^2} e^{-\pi/4(w/w_{\text{ave}})^2} \tag{3.23}$$

With Rayleigh distribution function, the CDF and the standard deviation are

$$F_{xy} = p(w_x \geq w \geq w_y) = e^{-\pi/4(w_x/w_{\text{ave}})^2} - e^{-\pi/4(w_y/w_{\text{ave}})^2} \tag{3.24}$$

$$\sigma = \sqrt{\left(\frac{4}{\pi} - 1\right) w_{\text{ave}}^2} \tag{3.25}$$

EXAMPLE 3.4

A wind turbine has wind blades of 24 m long. The average wind speed at the site is 6.9 m/s. The total efficiency of the system including C_p is 30%. Compute the annual energy produced assuming Rayleigh distribution for wind speed.

Solution:
The wind power equation is developed in Chapter 2.

$$P_w = \frac{1}{2} \delta A_{\text{blade}} w_u^3$$

The average power is

$$P_{w-\text{ave}} = \frac{1}{2} \delta A_{\text{blade}} (w_u^3)_{\text{ave}}$$

Modify Equation 3.11 to compute the average of the cube of wind speed

$$(w_u^3)_{\text{ave}} = \int_0^\infty w_u^3 f(w_u) dw_u = \int_0^\infty w_u^3 \left[\frac{\pi}{2} \frac{w_u}{w_{\text{ave}}^2} e^{-\pi/4(w_u/w_{\text{ave}})^2} \right] dw_u = \frac{6}{\pi} (w_{\text{ave}})^3$$

Hence,

$$P_{w-\text{ave}} = \frac{1}{2} \delta A_{\text{blade}} (w_u^3)_{\text{ave}} = \frac{3}{\pi} \delta A_{\text{blade}} (w_{\text{ave}})^3 = \frac{3}{\pi} (\pi \times 24^2) 6.9^3 = 567.66 \, \text{kW}$$

The average output electric power of the turbine is

$$P_{\text{out-ave}} = P_{w-\text{ave}} \eta = 567.66 \times 0.3 = 170.3 \text{ kW}$$

Annual energy production

$$E_{\text{annual}} = P_{\text{out-ave}} \times 8765 = 1.49 \text{ GWh}$$

3.4 Dependency and Repeatability

Meteorologists who specialize in predicting weather conditions divide the forecast into *synoptic* and *mesoscale*. The synoptic scale meteorology is regional forecast. It predicts variables such as air masses, fronts, and pressure for areas such as the northwest region of North America. The mesoscale meteorology forecasts local weather where topography, bodies of water, and urban heat island are often considered.

Because wind farms occupy a small region, synoptic forecast alone cannot be used to predict wind speed at a specific farm site. This is because it lacks sufficient local sampling and it does not include the impact of local topology (hill, forests, etc.). Mesoscale forecasting is what is needed for wind farms. Generally, it is often difficult to forecast wind with high accuracies because of several challenges:

- Regional data alone is not enough to forecast local conditions.
- Mesoscale forecasting requires sufficient number of local weather stations that are not available in most areas.
- Topography effects are hard to consider.
- For a given site, features that are directly impacting wind speed are not all known.

To improve on wind speed forecasting, two statistical variables are often used: correlation and cross-correlation. Correlation gives us the impact of a given feature on the local wind speed. Cross-correlation allows us to examine the repeatability of wind speed at the same site, or the delayed wind speed in one site downstream from a monitored site.

3.4.1 Cross-Correlation

Cross-correlation is a useful tool for wind power plants. It gives information on similarity between two wind speeds. For wind power plants, cross-correlation is important for the following two cases:

1. It identifies the relationship between two weather stations located in different wind power plants. This is particularly useful because wind speeds in neighboring wind power plants are often correlated: if they are at a distance from each other, one could predict the wind speed at one site based on the measurement made on the other site.

2. It can be used to evaluate the dependency of a given variable, such as air pressure or humidity on wind speed.

Assume that we have two measurements x and y. The cross-correlation coefficient of these two variables is

$$\rho_{xy} = \frac{E\left[(x - x_{ave})^T (y - y_{ave})\right]}{\sigma_x \sigma_y} \tag{3.26}$$

where:
ρ_{xy} is the cross-correlation coefficient between samples x and y
$E[.]$ is the expected value of "."
x_{ave} is the average value of sample x
y_{ave} is the average value of sample y
σ_x is the standard deviations of sample x
σ_y is the standard deviations of sample y

The cross-correlation coefficient is a good measure of the degree of similarity. Its value is in the range of −1 and +1. If the samples x and y are not correlated, ρ is zero. If they are closely related to each other, the magnitude of ρ is near 1. If it is positive, it indicates linear positive correlation (if one increases, the other increases as well and vice versa). When ρ is negative, the data has negative correlation (increase in one indicates a decrease in the other). When ρ equals ±1, it shows a perfect positive or negative fit.

EXAMPLE 3.5

Two sites with wind speeds as shown in the table. The number of samples is substantially reduced to allow for simple calculations without the use of spreadsheet.

w_x (m/s)	w_y (m/s)
4	1
6	3
8	6
10	8

Compute the correlation coefficient between the two sites.

Solution:
First step is to calculate the average of the two samples

$$w_{x-\text{ave}} = \frac{4+6+8+10}{4} = 7\,\text{m/s}$$

$$w_{y-\text{ave}} = \frac{1+3+6+8}{4} = 4.5\,\text{m/s}$$

Next, we need to calculate the standard deviations

$$\sigma_x = \sqrt{\frac{1}{4}\left[\sum_{i=1}^{4}(w_{xi} - w_{x-\text{ave}})^2\right]}$$

$$= \sqrt{\frac{1}{4}\left[\sum_{i=1}^{4}(4-7)^2 + (6-7)^2 + (8-7)^2 + (10-7)^2\right]}$$

$$= 3.0822\ \text{m/s}$$

$$\sigma_y = \sqrt{\frac{1}{4}\left[\sum_{i=1}^{4}(1-4.5)^2 + (3-4.5)^2 + (6-4.5)^2 + (8-4.5)^2\right]} = 2.6925\,\text{m/s}$$

The expected value is an averaging process

$$E\left[(w_x - w_{x-\text{ave}})^T(w_y - w_{y-\text{ave}})\right] = \frac{1}{4}\begin{bmatrix} 4-7 & 6-7 & 8-7 & 10-7 \end{bmatrix}\begin{bmatrix} 1-4.5 \\ 3-4.5 \\ 6-4.5 \\ 8-4.5 \end{bmatrix} = 6.0\,\text{m}^2/\text{s}^2$$

The cross-correlation coefficient is

$$\rho_{xy} = \frac{E\left[(w_x - w_{x-ave})^T(w_y - w_{y-ave})\right]}{\sigma_x \sigma_y} = \frac{6}{3.0822 \times 2.6925} = 0.7229$$

The data suggests a high positive correlation between the two sites. Therefore, it would be beneficial to use one site wind speed as a feature to forecast the wind speed of the other side.

Try to find a correlation coefficient between the sine and cosine waves for sampling up to 90°, 180°, and 360°. Why the correlation coefficient changes with sample range?

For continuous function, the cross-correlation coefficient can be expressed by

$$\rho_{xy} = \int_{-\infty}^{\infty} x(t)y(t)dt \tag{3.27}$$

where:
  $x(t)$ is the function of variable x
  $y(t)$ is the function of variable y

For two sites with time lag correlation, the cross-correlation function is

$$\rho_{xy} = \int_{-\infty}^{\infty} x(t)y(t-\tau)dt \tag{3.28}$$

where:
  τ is the time lag between the two sites

3.4.2 Repeatability

Cross-correlation can also be used to provide a degree of similarity between a given time series and a lagged version of itself over successive time intervals. Patterns in wind speeds may exist, which could give important information on the repeatability of wind speeds over time. Figure 3.6 shows a wind speed pattern at a given site, which is changing with time.

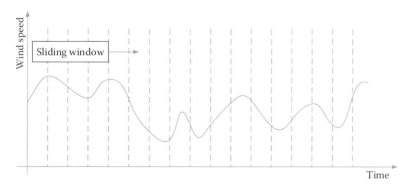

FIGURE 3.6
Sliding window.

If a window of time is selected, we can slide it over the entire time range and sample wind speed as we go.

The cross-correlation between two samples in two different windows is the same as the correlation coefficient discussed earlier.

$$R = \frac{E\left[(w_1 - w_{1-\text{ave}})^T (w_2 - w_{2-\text{ave}})\right]}{\sigma_1 \sigma_2} \tag{3.29}$$

where:
  R is the cross-correlation coefficient
  w_1 is the wind speed vector for first window
  w_2 is the wind speed vector for second window
  $w_{1-\text{ave}}$ is the average wind speed of first window
  $w_{2-\text{ave}}$ is the average wind speed of second window

For continuous function, the cross-correlation can be expressed by

$$R(k) = \int_{-\infty}^{\infty} f(t) f(t+k) dt \tag{3.30}$$

where:
  k is the width of the moving window
  $f(t)$ is the function to be evaluated

Exercise

1. What is the difference between the frequency and probability of occurrence?
2. What is the variance of data?
3. What is the standard deviation of data?
4. What is the cumulative distribution function?
5. What is the probability density function?
6. What is the difference between Weibull and Rayleigh distribution functions?
7. Why normal distribution is not suitable for wind applications?
8. What is dependency?
9. What is repeatability?
10. How does dependency and repeatability improve wind forecasting?
11. The following measurements were made at a potential site for wind farm located at 100 m elevation:

w_i (m/s)	n_i (h)
0	400
2	600
4	700

(Continued)

w_i (m/s)	n_i (h)
6	800
8	900
10	1000
12	1200
14	1100
16	900
18	800
20	200
22	100
24	50
26	15

Compute the average wind speed and the standard deviation. If the turbines' cut-in speed is 4 m/s and the cut-out speed is 16 m/s, compute the percentage of time the turbine can generate electricity.

12. A Weibull function representing wind speed at a given site has $c = 3$ and $k = 0.5$. Compute the average wind speed.

13. A Weibull function representing wind speed at a given site has $c = 10$ and $k = 2.17$. Compute the number of hours per year when wind speed is 9 m/s or greater. Assume 8760 h/year.

14. The wind speed and its frequency for a site is given in the table below. Find Wiebull and Rayleigh distribution function for wind PDF.

w_i (m/s)	n_i (h)
1	70
2	130
3	190
4	230
5	300
6	320
7	330
8	330
9	310
10	300
11	270
12	250
13	230
14	200
15	180
16	160
17	130
18	100
19	90
20	70

15. A wind turbine has wind blades of 50 m long. The total efficiency of the system including C_p is 30%. Compute the annual energy produced assuming Weibull distribution for wind speed. Assume $c = 1$ and $k = 1$ for the distribution function. Assume that the cut-in speed is 4 m/s and cut-out speed is 20 m/s.

16. The average wind speed at the site is 6.9 m/s. Compute the average wind power density assuming Rayleigh distribution for wind speed.

17. A wind turbine has 50 m tower. The turbine is installed offshore in calm water, where the average wind speed at 10 m height is 5 m/s. Compute the wind speed at the hub level.

18. A Wiebull function representing wind speed at a given site has $c = 5$ and $k = 1.2$. Compute the annual average wind speed.

19. A Wiebull function representing wind speed at a given site has $c = 5$ and $k = 1.2$. Compute the number of hours per year when wind speed is 4 m/s or greater.

20. A wind turbine has wind blades of 50 m long. At an upwind speed of 15 m/s, the blades rotate at 30 r/min. Assume that the wind speed increases by 10%, and the pitch angle is adjusted to keep the tip-speed ratio constant. Compute the new rotating speed of the blade.

21. Two sites with wind speeds given in the table. Compute the cross-correlation of the wind speed of the two sites.

Time (h)	w_1 (m/s)	w_2 (m/s)
0.0	9.000	2.951
2.0	11.114	4.809
4.0	12.524	6.943
6.0	13.143	8.996
8.0	13.020	10.665
10.0	12.309	11.740
12.0	11.224	12.125
14.0	9.998	11.839
16.0	8.843	11.002
18.0	7.921	9.802
20.0	7.330	8.465
22.0	7.099	7.208
24.0	7.200	6.213
26.0	7.555	5.603
28.0	8.066	5.426
30.0	8.626	5.661
32.0	9.141	6.225
34.0	9.542	6.995
36.0	9.787	7.831
38.0	9.869	8.598
40.0	9.803	9.187
42.0	9.628	9.530
44.0	9.389	9.600
46.0	9.134	9.420
48.0	8.905	9.044

22. A wind site has wind speed as given in the table. Compute the cross-correlation between the two consecutive days.

Time (h)	w (m/s)
2.0	11.290
4.0	13.135
6.0	14.267
8.0	14.536
10.0	13.937
12.0	12.594
14.0	10.747
16.0	8.702
18.0	6.782
20.0	5.282
22.0	4.418
24.0	4.298
26.0	4.913
28.0	6.137
30.0	7.758
32.0	9.508
34.0	11.110
36.0	12.322
38.0	12.972
40.0	12.979
42.0	12.369
44.0	11.260
46.0	9.844
48.0	8.353

4

Overview of Wind Turbines

Until the middle of the twentieth century, wind turbine designs failed to provide the reliability and efficiency needed for serious consideration as electric power producers. However, as with any technology, wind turbines have evolved over the years and the major innovations were made from 1990s. With the advancement in material, structure, power electronics, and control, we have now wind turbines that are proven reliable, efficient, and worthy of grid integration. These turbines are also cost effective and can produce electricity at rates comparable to conventional thermal generation. The present size of wind turbines ranges from just a few kW to 8 MW. The size is continuously increasing and we should expect 10–15 MW within a decade or two.

Wind turbines come in a plethora of designs with different generators, configurations, and control strategies. The most common generators used in wind turbines are the squirrel-cage induction generator (SCIG), slip-ring induction generator (SRIG), synchronous generator (SG), and permanent magnet SG (PMSG). The basic system, called "type 1," has little or no control on its generator. The other systems, types 2–4, use power electronic to provide various control actions. In this chapter, the configurations of the most common types of wind energy system are discussed.

4.1 Classification of Wind Turbines

Nowadays, wind turbines have several designs with a plethora of features. To classify wind turbines, engineers use features such as the alignment of the rotating axis, type of electrical generator, speed of rotation, power conversion, and control actions. For utility size turbines, the industry has established a type system to describe the general design and features of wind turbines. In this section, the classification based on features is discussed, and in Section 4.2, the types of wind turbines are given.

4.1.1 Alignment of Rotating Axis

The drive shafts of wind turbines can rotate horizontally or vertically. A horizontal-axis wind turbine (HAWT) is shown in Figure 4.1. This is the most common type of wind turbine used today. Its main drive shaft, gearbox, electrical generator, brakes, actuators, and the transformer in some designs are housed inside a nacelle at the top of a tower. The basic components of the HAWT (shown in Figure 4.2) are as follows:

- A tower that keeps the rotating blades at a sufficient height to increase the exposure of the blades to faster wind. Large wind turbines can have their towers as high as 250 m above the base. The tower withstands tremendous sheer and bending forces, so its structure is made of steel.

- Rotating blades that capture the kinetic energy of the wind. They are normally made of material such as fiberglass-reinforced polyester or wood epoxy. The length of the rotating blades ranges from 5 to over 100 m. To strengthen its structure, the blade is currently manufactured as a single solid unit without any sectionalization. However, newer research indicates that sectionalizing the blade could be possible. If done, it would allow the transportation and installation of large turbines on land.

- A hub that is connected to the low speed shaft of the gearbox. The blades are mounted on the hub and their pitch angle can be adjusted by actuators.

- A yaw mechanism that rotates the nacelle to face the upwind. Thus, increasing the exposure of the blades to wind.

- A gearbox that connects the low-speed rotating blades to the high-speed generator.

- A generator that is mounted on the high-speed shaft of the gearbox to convert the mechanical energy of the rotating blades into electrical energy.

- A disk brake to prevent the blades from rotating when wind conditions are not suitable for generating electricity.

- A transformer that steps up the output voltage of the generator.

- Several controllers that integrate the wind turbine into the utility grid and regulate the generated power. They also protect the turbine against severe conditions such as grid faults and wind storms.

For large turbines, the blades are placed at a distance in front of the tower and tilted up a little. This is done to prevent the blades from hitting the tower at high wind conditions as well as at rapid pitching of the blades.

FIGURE 4.1
Horizontal-axis wind turbine system.

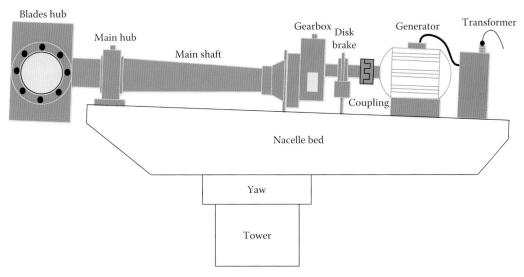

FIGURE 4.2
Mechanical structure of a HAWT.

The main advantages of the HAWT are

- The tall tower allows the turbine to access strong wind.
- It is a high-efficiency turbine as its blades continuously receive power from wind during the entire rotation.
- The speed of the blade is fairly constant during a single rotation. Thus, rapid fluctuations in electrical variables such as voltage and reactive power are insignificant.

The main disadvantages of HAWT are as follows:

- It requires massive tower construction to support the heavy equipment in the nacelle.
- The heavy generator, gearbox, and transformer inside the nacelle have to be lifted during construction and maintenance.
- It requires an additional yaw control system to turn the blades toward wind.

The vertical-axis wind turbine (VAWT) is shown in Figure 4.3. It is known as "Darrieus wind turbine" and it looks like a giant eggbeater. It has basically the same components, but without a yaw mechanism. The main advantages of the VAWT are as follows:

- The generator, gearbox, and transformers are all located at the ground level, making it easier to install and maintain them as compared to the HAWT.
- There is no need for a yaw mechanism to direct the blade into wind. This is an advantage for sites with variable wind directions.
- The cut-in speed of the VAWT is generally lower than that for the HAWT.

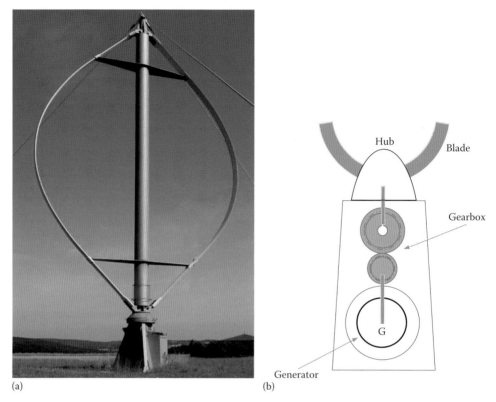

(a)   (b)

FIGURE 4.3
Vertical-axis wind turbine: (a) horizontal design and (b) mechanical structure. (Courtesy of US National Renewable Energy Lab.)

The main disadvantages of the VAWT are as follows:

- The wind speed is slower near the ground. Hence, the available wind power for VAWT is less than that for HAWT.
- Air flow near the ground and other objects can create turbulent flow. This can introduce mechanical vibrations in the turbine components that eventually shorten the lifetime of the turbine.
- Because of its massive inertia, they may require external power source to startup the turbine.
- The bearing at the base carry the heavy load of the blades. Their failure rate is high.

4.1.2 Types of Generators

Most wind turbines utilize asynchronous (induction) and synchronous machines. These two types of machines are covered in Chapters 6 and 7. The common generator used today in wind turbines is the induction generator, which has two types: *squirrel cage* and *wound rotor (slip ring)*. The squirrel-cage machine has no access to its rotor circuit. Hence, it is the cheapest and the most rugged machine. However, there is little control that can be implemented on these machines. The wound rotor induction generator has access to its rotor circuit through a mechanism of brushes and slip rings. This allows us to do two main control functions: inject a voltage signal into the rotor or insert external resistance in the rotor circuit. Doing any of these actions, we can have control on the system performance, as discussed in Chapters 9 and 10.

The induction generator operates at a slip so its speed is not exactly constant. The machine itself without rotor injection cannot generate electricity unless it is rotating at higher than its synchronous speed n_s

$$n_s = 120 \frac{f}{p} \tag{4.1}$$

where:
  n_s is the synchronous speed in rpm
  f is the frequency of the grid in Hz
  p is the number of poles of the machine

The slip of the machine s is defined as

$$s = \frac{n_s - n}{n_s} \tag{4.2}$$

The slip of the induction generator, without rotor injection, is often small, 2%–10%, and the generator must have a negative slip to produce electricity. This will be covered in detail later in Chapters 6 and 8.

The SG is divided into two types: electric magnet and permanent magnet. For both types, the frequency of the power produced is directly proportional to the speed of its rotation. There is no slip in these machines. Because the rotation of the generator shaft is varying due to variations in wind speed, the output frequency of these machines is not constant. Hence, a converter between the generator and the grid is needed to deliver electricity at the grid frequency.

The electric magnet machine allows access to its field circuit. By controlling the field current, the voltage and reactive power at the terminals of the machine can be controlled. The permanent magnet machines do not have field current control and are often made out of rare earth magnetic material to provide strong electric field.

A newer type of wind turbines utilizes the synchronous machine without gearbox. This type of generator has a large number of poles to reduce its synchronous speed and allow for power generation at low blade speed. Because of the large number of poles, the diameter of this machine is quite large making the nacelle very wide.

4.1.3 Speed of Rotation

Wind turbines can be divided into two classes: fixed-speed wind turbine (FSWT) and variable-speed wind turbine (VSWT). The older systems were mostly fixed-speed type, as they are simpler to build and operate than the variable-speed types. They generate electricity only when the wind speed is high enough to spin the generator shaft above its synchronous speed. Although they are less expensive and require less maintenance than the variable-speed types, they are limited in their power generation.

4.1.3.1 Fixed-Speed Wind Turbine

Figure 4.4 shows the torque-speed characteristic of the FSWT. The details of the machine are given in Chapter 6. The generator operates in almost linear range from point 1 to point 2, as shown in the figure. Before 1, the speed of the turbine is below the cut-in speed. When the speed is higher than that at point 2, the machine enters into its unstable nonlinear region. Because the steep slope of the characteristic, the range of speed is very narrow, thus it is called "constant" or "fixed-speed turbine." The range of the developed power (shaft power of the generator) in this system is

$$\Delta P_{cs} = P_2 - P_1 = T_2 \omega_2 - T_1 \omega_1 \tag{4.3}$$

where:
ΔP_{cs} is the range of the developed power of constant speed generator
P_1 is the developed power at point 1
P_2 is the developed power at point 2
T_1 is the developed torque at point 1
T_2 is the developed torque at point 2
ω_1 is the speed of the generator at point 1
ω_2 is the ω speed of the generator at point 2

Because

$$\omega_2 \approx \omega_1 \tag{4.4}$$

then,

$$\Delta P_{cs} = \left(T_2 - T_1 \right) \omega_1 \tag{4.5}$$

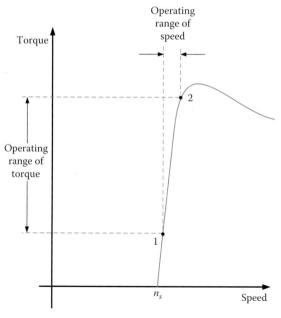

FIGURE 4.4
Torque-speed characteristics of a fixed-speed wind turbine.

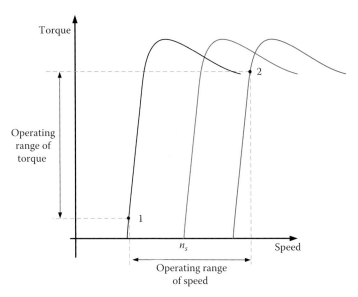

FIGURE 4.5
Torque-speed characteristics of variable-speed wind turbine.

Because the speed of the FSWT is fairly constant, the output power is regulated by adjusting the lift force using the pitch angle control.

4.1.3.2 Variable-Speed Wind Turbine

The VSWTs are more complex systems with various power electronic converters that allow the generators to produce electricity at a wide range of speeds, even at less than synchronous speeds. Thus, their range of operation is much wider than that for the FSWT. The torque-speed characteristics of the VSWT are shown in Figure 4.5. To achieve these characteristics, a voltage is injected in the rotor circuit of the generator.

If the operating torque is between T_1 and T_2, the speed of the generator has a wide range as $\omega_2 > \omega_1$. The range of power for this machine is

$$\Delta P_{vs} = P_2 - P_1 = T_2\omega_2 - T_1\omega_1 \tag{4.6}$$

where:
ΔP_{vs} is the developed power of the variable-speed generator

Because the wider variation of operating speed, $\Delta P_{vs} > \Delta P_{cs}$.

EXAMPLE 4.1

A fixed-speed generator has a torque range from 500 to 3000 NM. The synchronous speed of the generator is 1200 rpm. Estimate the range of the developed power of the generator assuming the slip is −0.02.

Solution:
Using Equation 4.4 for fixed-speed generator, and Equation 4.2 for the slip, we can compute the speed of the generator

$$\omega_1 \approx \omega_2 = \omega_s(1-s) = 2\pi\frac{1200}{60}(1+0.02) = 128.18 \text{ rad/s}$$

The range of the developed power is

$$\Delta P_{cs} = (T_2 - T_1)\omega_1 = (3000 - 500) \times 128.18 = 320.45 \text{ kW}$$

EXAMPLE 4.2

Assume that the generator in the previous example is equipped with converter that can operate the generator at a speed range of 900–1500 r/min. Estimate the range of the developed power.

Solution:
The range of the developed power using Equation 4.6 is

$$\Delta P_{vs} = T_2\omega_2 - T_1\omega_1 = 3000 \times 2\pi\frac{1500}{60} - 500 \times 2\pi\frac{900}{60} = 424.12 \text{ kW}$$

This is about a 32% increase in power range over the fixed-speed system.

4.1.3.3 Assessment of FSWT and VSWT

Figure 4.6 shows the generic power–speed characteristics of constant and variable-speed turbines. For fixed-speed turbine, the cut-in speed (minimum speed for generating electricity) is higher than the synchronous speed of the generator, and its cut-out speed is determined by the maximum speed of the turbine at point 2 in Figure 4.4. For VSWT, the cut-in speed is lower than the synchronous speed, and the cut-out speed is higher than that for FSWT, as shown in Figure 4.5. For FSWT, the rated power (which is the power delivered by its stator windings) is less than that for VSWT. For VSWTs, the output power is delivered by the stator and rotor. The stator can still provide its rated power and the extra is provided through the rotor. The details of these analyses are given in Chapter 10.

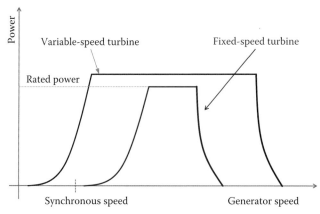

FIGURE 4.6
Power generation for fixed- and variable-speed wind turbines.

Although it is relatively primitive design, FSWT is still used for small size systems because of several reasons; some of them are as follows:

- Does not require brushes or slip rings
- Low maintenance
- Rugged generator
- Low cost
- Simple to operate

The FSWT has several disadvantages, of which some are as follows:

- Because of the fixed-speed operation, fluctuations in wind speed as well as gusts translate into continuous and sudden torsional torques that stress the drive train shaft and gearbox.
- The speed of rotation that generates electricity is above the synchronous speed, which is high. This requires gears with high ratios or blades that rotate at high speeds. In either case, the tower is built for higher structural loads.
- Because of the high speed, the FSWT could be noisy and could cause more bird collisions.

VSWT is much more popular nowadays because of its overwhelming advantages:

- Can produce power at low speeds (lower than the synchronous speed)
- Output power can be regulated even when the speed of the turbine changes widely
- Speed of the generator can be adjusted to achieve higher aerodynamic efficiency (maximize the coefficient of performance)
- Lower mechanical stress due to the reduction of the drive train torque variations
- Noise and bird collision problems are much reduced because the turbine operates at low speeds

The main drawbacks of the VSWT are as follows:

- High cost
- More complex system
- More components are used that increases the maintenance cost

4.1.4 Power Conversion

Depending on the type of generator, its output terminals can either be directly or indirectly connected to the grid. For asynchronous machine (induction generator), the connection to the grid is direct without any extensive conversion, except for soft starting, as seen in Chapter 8. Because of the slip, the frequency of the output voltage is locked to the frequency of the grid even when the rotor speed varies.

For SG, the frequency of its terminal voltage is directly dependent on the speed of rotation. Because the wind speed is varying, the frequency of the terminal voltage varies as well. If a SG is directly connected to the grid and is producing energy at different frequency than the grid frequency, the generator will be tripped off the grid to protect the generator from destructive torsional oscillations on its shaft as well as damaging excessive current in its windings. Therefore, a power converter is installed between the generator and the grid. The converter, which is an ac/ac type, is designed to convert the variable frequency output of the generator to the fixed frequency of the grid. The converter is often called "frequency converter" or "full converter."

4.1.5 Control Actions

Wind turbines and wind farms have several types of controls. The most common ones are soft starting, generation control, pitch control, feathering, reactive power control, stability, low- and high-voltage ride through, and ramp control.

4.1.5.1 Soft Starting

Wind turbines are often connected to the grid through a soft starting mechanism to reduce the initial transients such as inrush current, voltage dips, and mechanical stress. The soft starting mechanism is often made of voltage control converter connected between the generator and the grid. It allows the voltage on the terminals of the generator to ramp up at a rate that would not create unacceptable transients. When the rotation of the turbine reaches the operating speed, the generator is connected to the grid through the soft starter. After the voltage across the generator reaches the grid voltage, the soft starter is removed from the circuit.

4.1.5.2 Generation Control

With generation control, the output power of the turbine, or the wind farm, is adjusted based on contractual agreement with the serving utility. This is one of the automatic generation control (AGC) functions. AGC can also provide adjustable generation to compensate for variations in demand. Of course, the generation control is only effective if the right wind conditions exist.

Pitch angle adjustments are used to control the output power of the turbine. Its control depends on the region of operation of the wind turbine, which is shown in Figure 4.7. When wind speed is in region 1, the generator is disconnected from the grid as wind speed is lower than needed to generating electricity. When wind speed is in region 2 (higher than

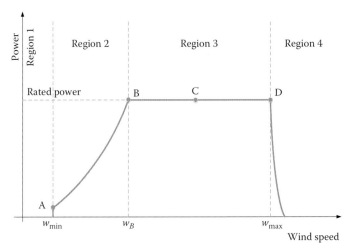

FIGURE 4.7
Regions of wind power generation.

w_{min} in the figure), the pitch angle is adjusted to maximize the output power by tracking the maximum C_p, as given in Chapters 8 through 12. When wind speed is in region 3 (when the output of the generator is at its rated value), the pitch angle is adjusted to spill some of the wind to prevent the machine from overcurrent damages. When wind speed exceeds the cut-out speed (damaging wind speed) in region 4, the blades are adjusted to minimize the lift force and disk brake is applied. This is known as "feathering."

4.1.5.3 Reactive Power Control

Most of the induction generators have no excitation circuits. Thus, the reactive power comes from the grid. Induction generator consumes large amount of reactive power which can cause voltage depression at the farm. To solve this problem, external sources such as adaptive reactive power compensators are used. Type 3 wind turbines are equipped with excitation circuit that can help meet the reactive power demand.

4.1.5.4 Stability Control

To ensure power system stability, it is important to match all generations with all demands at all times. The demands include the consumption of energy by customers in addition to system losses. If the balance between generations and demands is not maintained, blackout could occur. Also, sudden change in demand or generation can cause the power system to oscillate. If not damped quickly, the oscillation can cause system outage. Also, if a generator is tripped (disconnected) during a temporary fault, the grid becomes deficient in generation after the fault is cleared. This leads to a forced outage to balance demand to generation.

   To avoid this problem, as discussed in Chapter 12, wind turbines are required to provide the following two functions:

* Control its output power to help maintain grid stability.
* Stay connected to the grid during temporary faults. This is known as "low voltage ride through."

4.1.5.5 Ramping Control

Ramping is another important control function for wind turbines. Ramping up the electric power either up or down is important during steady-state and transient operations. During steady state, utilities may require wind farms to ramp up or down slowly to allow the system to modify the rest of its generations to maintain the energy balance and prevent unstable operations. However, during transient operations, such as temporary faults, utilities may require the wind turbines to ramp quickly to compensate for energy deficit or surplus to prevent unnecessary outages or unstable operation.

4.2 Types of Wind Turbines

Wind turbines have various designs and configurations. For large wind power turbines, engineers classify them into five types according to their components and the modes of their operations. Although the types may have specific design and capabilities, variations within a given type are possible. The detailed operation and modeling of the wind turbine types are given in Chapters 8 through 11.

4.2.1 Type 1 Wind Turbine

Type 1 wind turbine is a fixed-speed system, whose main components are shown in Figure 4.8. It consists of blades mounted on a hub (also called "turbine rotor") and are capable of pitch angle control. The shaft of the hub is connected to a gearbox to step up the speed of the generator. This is because the speed of the blade is just a few revolutions per minute, which is way below the speed at which the generator operates. The high-speed shaft of the gearbox is connected to a SCIG. The rotor circuit of the squirrel-cage generator is enclosed on itself and cannot be accessed. Hence, the generator has limited control capabilities. The speed is controlled by the pitch angle of the blade, and the reactive power can only be controlled by external devices such as switching capacitors or other dynamic reactive power compensators. When wind speed is favorable, the generator is connected to the grid through a soft starter to reduce the initial voltage at the terminals of the generator, thus limiting the inrush current, which is mainly reactive current. The soft starter is an ac/ac voltage converter that is bypassed after the generator starts, which is discussed in Chapter 8.

The turbines in the farm are connected to a common bus called "farm collection bus." The voltage of this bus is lower than that for the connecting transmission line; the output voltage of the turbine is often 690 V. To step up this voltage to the grid voltage, a generation step-up (GSU) transformer (xfm) is used. If the wind farm is located away from the grid, a transmission line (often called "trunk line") is used to connect the farm to the grid.

The characteristic of type 1 system is shown in Figure 4.4. As seen in the figure, the range of operating speed is very narrow. This is because the slope is dependent on the rotor resistance, which is quite small for large squirrel-cage machines. The detailed operation of type 1 system is given in Chapter 8.

Most of the cost of wind turbines is due to a few of its key components. The turbine rotor accounts for approximately 20% of the total cost. The generator and gearbox account for about 35% of the cost. The structural support account for about 15% of the turbine cost.

4.2.2 Type 2 Wind Turbine

Type 2 wind turbine has basically the same components as those for type 1 with one exception; the generator is wound rotor induction machine. This generator has its rotor circuit accessible by external circuits. Instead of shorting the rotor winding as with the squirrel-cage generator, the rotor windings are connected to a switching device that insert a resistance, as shown in Figure 4.9. The power consumed by the resistance is regulated by the duty ratio of the switching circuit. In electric drives, this operation is well known and is called "dynamic braking." The function of the resistance is to consume excess power.

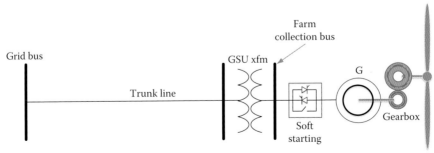

FIGURE 4.8
Type 1 wind turbine system.

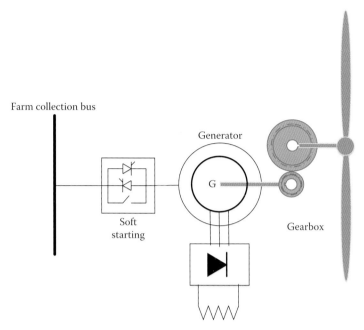

FIGURE 4.9
Type 2 wind turbine system.

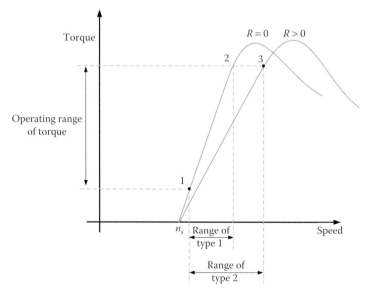

FIGURE 4.10
Typical torque-speed characteristics of type 2 wind turbine generator.

But why do we need to do that for wind turbines? The reason is that the machine can operate at wider range of speed as compared with type 1 systems. Also, it regulates the delivered power to the grid. This method, of course, is low efficiency.

The torque-speed characteristics of this type of wind turbine are shown in Figure 4.10. The figure shows the characteristics of the generator with and without the added resistance to the rotor circuit. For the same range of operating torque, the range of the operating speed increases for type 2 when the resistance is added. Thus, the turbine can produce power at a wider range of wind speed as compared with type 1. However, because the increase in the speed range is often modest, type 2 turbine is called "variable slip system," where the slip can be adjusted by just 10%. The term "variable speed" is often reserved for types 3–5 systems.

Another advantage for this system is its soft starting as the external resistance reduces the inrush current during starting. In a newer design, the switching circuit is embedded into the rotor of the generator and there is no need to use brushes and slip rings, which are high maintenance components. On the down side, type 2 turbines have no internal reactive power compensation. Therefore, external compensation devices are needed. The detailed operation of type 2 turbines is given in Chapter 9.

4.2.3 Type 3 Wind Turbine

Type 3 is probably the most widely used system to date. Its main components are shown in Figure 4.11. It consists of a wound rotor induction generator with the rotor connected to the grid through two converters: rotor-side converter (RSC) and grid-side converter (GSC). Because the terminals of the generator as well as the rotor are connected to the grid, it is called "doubly fed induction generator (DFIG)." The bus between the two converters is direct current. Each of these converters operates in ac/dc or dc/ac modes. The GSC links the three-phase terminal voltage of the generator to the dc bus. The RSC links the three-phase windings of the rotor to the dc bus. The detailed operation of the DFIG is given in Chapter 10.

DFIG has several advantages, of which some of them are as follows:

- It has a much wider operating speed range as compared to either types 1 or 2. The slip can change by as much as 50% as compared to 10% for type 2. This allows the turbine to capture more energy from wind. The characteristics of the DFIG are shown in Figure 4.5.

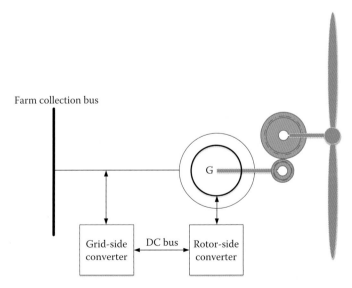

FIGURE 4.11
Type 3 wind turbine system (DFIG).

- Power flow through the rotor is bidirectional allowing the generator to operate at sub- and super-synchronous speeds.
- Reactive power can be controlled and compensated by the converters.
- Stability of the farm can be enhanced.
- Converters process a maximum of 30% of the rating of the generator. This makes the power electronic circuit relatively small as compared with type 4.

4.2.4 Type 4 Wind Turbine

One of the main components of wind turbines in types 1–3 is the gearbox, which is needed to run the generator at high enough speed for energy generation. However, the gearbox is expensive, heavy, and one of the main failure modes of the system. Type 4 wind turbines can operate without the gearbox, as shown in the common design in Figure 4.12. It consists of blades that are connected directly to a synchronous generator. The terminals of the generator are connected to the grid through an ac/ac converter.

To understand the operation of this system, keep in mind that the SG produces ac power at a frequency proportional to its rotating speed, as given in Equation 4.1. If the speed of the generator is n, the frequency of the generated power is

$$f = \frac{n}{120}p \qquad (4.7)$$

Because the speed of the generator can vary based on wind conditions, the frequency of the generated electricity varies as well. Because the frequency of the grid and that of the generator output must be identical to prevent the tripping of the generator, the turbine cannot be connected directly to the grid without matching the grid frequency. This is the function of the ac/ac converter.

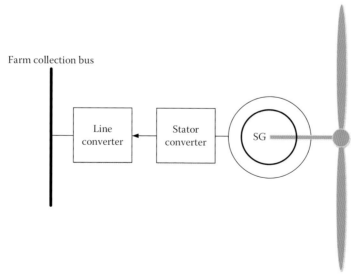

FIGURE 4.12
Type 4 wind turbine system.

Because we have no gearbox, the speed of the generator is very low. Hence, the generator of type 4 is made of a high number of poles similar to the generators used in hydroelectric power plants (often greater than 40 poles). This way, the output frequency of the generator is closer to the grid frequency at low speeds.

The converters in type 4 system processes the entire output power of the generator, its steady-state rating is at least equal to the rating of the generator itself. Thus, it is called "full converter." Its cost is much higher than the cost of the converter in type 3 system.

Type 4 wind system has several advantages, of which some of them are as follows:

- Operates at the widest speed range among all other types; speed variation from almost zero to above synchronous speed is possible.

- Reactive power can be easily controlled by the excitation of the generator, or through the converter.

- Power engineers are very familiar with the synchronous generator, which is used in all conventional power plants. This makes actions such as generation control, protection, and stability analysis very similar to well-known technologies.

Type 4 system has some limitations. For instance, its generator is very large in diameter because of its large number of poles, and is more expensive to build and install. Although the absence of the gearbox is advantageous from the cost and maintenance viewpoint, its removal reduces the overall inertia of the system. This inertia is very important in power system operation because it acts as a temporary storage of energy during fluctuations and disturbances, as discussed in Chapter 12.

4.2.5 Type 5 Wind Turbine

Type 5 wind turbine system is shown in Figure 4.13. It consists of a multistage gear system connected to a torque/speed converter. This mechanical coupling system is connected to a synchronous generator.

The torque/speed converter is similar to the one used in automatic transmission vehicles. It is a type of fluid coupling, which allows the generator to spin at a speed that is independent of the speed of the gearbox. Thus, the generator is not rigidly connected to the gearbox, and the generator speed can be maintained constant even when the output speed of the gearbox varies.

This coupling is equivalent of a gear system with continuous ratio. The multistage gearbox is used to select the proper ratio for the existing wind speed. The ratio rotates the output shaft of the gearbox at a speed higher than the synchronous speed of the generator. The torque/speed converter reduces the speed to the synchronous speed of the generator. With this system, the generator is always rotating at the synchronous speed, and it can be directly connected to the grid without the use of any electronic converter. This system is similar to conventional generation if you assume that the gearbox and torque/speed converter act as the governor in hydroelectric power plant or steam values in thermal power plants.

The generator for this system has small number of poles, typically 6 poles or 4 poles. This is similar to the generators used in thermal power plants.

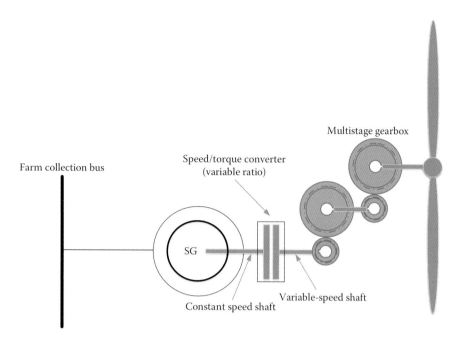

FIGURE 4.13
Type 5 wind turbine system.

Exercise

1. How are wind turbines classified?
2. What are the advantages and disadvantages of HAWT?
3. What are the advantages and disadvantages of VAWT?
4. What is the function of the hub?
5. What is the function of the yaw?
6. What is the function of the gearbox?
7. What are the various types of converters used in types 2–5?
8. What are the main control actions in wind power plants?
9. State the advantages and disadvantages of type 1 wind turbine.
10. State the advantages and disadvantages of type 2 wind turbine.
11. State the advantages and disadvantages of type 3 wind turbine.
12. State the advantages and disadvantages of type 4 wind turbine.
13. State the advantages and disadvantages of type 5 wind turbine.
14. For type 1 system, how is the generator excited?
15. Can you control the excitation of the permanent magnet synchronous generator?
16. What are the main functions of rotor-side converter?

17. What are the main functions of grid-side converter?
18. What are the main types of generators used in wind energy?
19. What are the advantages and disadvantages of fixed-speed wind turbines?
20. What are the advantages and disadvantages of variable-speed wind turbines?
21. What is soft starting?
22. Can induction generator produce reactive power on its own?

5

Solid-State Converters

Power electronics play major roles in wind energy. They are responsible for several functions such as starting the system, regulating turbine speed, interfacing turbine with the grid, regulating real and reactive powers, and controlling the system during disturbances. Throughout the last three decades, the developments of power electronic circuits were remarkable as several sophisticated systems were designed and used in wind energy. It is because of these systems that we have types 2, 3, and 4.

Solid-state devices are the main building components of any converter. Their function is mainly to mimic the mechanical switches by connecting and disconnecting electric loads, but at very high speeds. Unlike other applications of electronics, power electronic devices are either fully closed or fully open. Operating in any other mode for even a short time can destroy the devices.

Solid-state switching circuits (converters) have four types, as shown in Figure 5.1. The ac/dc converter converts any ac waveform into a dc waveform with adjustable voltage or current. The dc/dc converter converts a dc waveform into an adjustable dc waveform. The dc/ac converter converts a dc waveform into an ac waveform with adjustable voltage and frequency. The ac/ac converter converts a fixed voltage, fixed frequency ac waveform into an adjustable ac waveform in terms of voltage and frequency.

5.1 AC/DC Converters with Resistive Load

The ac/dc converter is used to produce a dc waveform from an ac source. The output dc waveform is either fixed or variable voltages depending on the design. The simplest form of ac/dc converter is the rectifier circuit. More elaborate circuits use switching device such as the bipolar junction transistor, field effect transistor (FET), metal–oxide semiconductor FET, insulate gate bipolar transistor, and silicon-controlled rectifier (SCR) to control the dc voltage or current.

5.1.1 Rectifier Circuits

A simple full-wave ac/dc rectifier circuit is shown in Figure 5.2. The circuit consists of an ac source of potential v_s, a load resistance, and four diodes in bridge configuration. The diode allows the current to flow in one direction only when it is forward-biased. The circuit in Figure 5.2a represents the case when the source voltage is in the positive half cycle; point A has higher potential than point B. Hence, diodes D_1 and D_2 are forward-biased and the current flows as shown in Figure 5.2a. The circuit in Figure 5.2b is for the negative half of the ac cycle; point B has higher potential than point A. In this case, D_3 and D_4 are forward-biased and the current flows as shown in Figure 5.2b. In either half of the ac cycle, the current in the load is unidirectional, and the voltage is always positive. The waveforms of the circuit are shown in Figure 5.3.

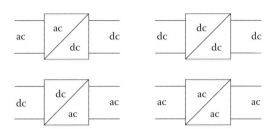

FIGURE 5.1
Four types of converters.

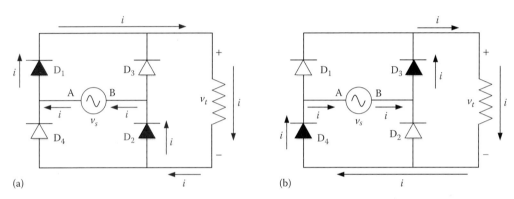

FIGURE 5.2
Full-wave rectifier circuit: (a) $v_{AB} > 0$ and (b) $v_{AB} < 0$.

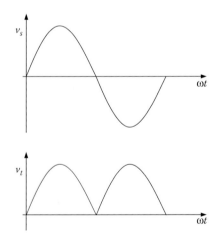

FIGURE 5.3
Waveforms of full-wave rectifier circuit.

Let us assume that the source voltage is sinusoidal

$$v_s = V_{max} \sin \omega t \tag{5.1}$$

The average voltage across the load can be obtained from the waveforms in Figure 5.3

$$V_{ave} = \frac{1}{2\pi} \int_0^{2\pi} v_t \, d\omega t = \frac{1}{\pi} \int_0^{\pi} v_s \, d\omega t = \frac{V_{max}}{\pi} \int_0^{\pi} \sin \omega t \, d\omega t = \frac{2V_{max}}{\pi} \tag{5.2}$$

In root mean square (rms) representation, the rms voltage of the load is defined as

$$V_{rms} = \sqrt{\frac{1}{2\pi} \int_0^{2\pi} v_t^2 \, d\omega t} \tag{5.3}$$

$$V_{rms} = \sqrt{\frac{V_{max}^2}{\pi} \int_0^{\pi} \sin^2 \omega t \, d\omega t} = \frac{V_{max}}{\sqrt{2}}$$

Notice that the rms voltage across the load for the full-wave rectifier circuit is the same as that for the source voltage. The rms current of the load is

$$I_{rms} = \frac{V_{rms}}{R} \tag{5.4}$$

where:
  R is the load resistance

The electric power consumed by the resistive load is

$$P = \frac{V_{rms}^2}{R} = R I_{rms}^2 \tag{5.5}$$

EXAMPLE 5.1

A full-wave rectifier circuit converts a 120 V (rms) source into dc. The load of the circuit is a 10 Ω resistance. Compute the following:

1. Average voltage across the load
2. Average voltage of the source
3. Voltage of the load in rms
4. Current of the load in rms
5. Power consumed by the load

Solution:

1. $V_{ave} = \dfrac{2V_{max}}{\pi} = \dfrac{2\left(\sqrt{2} \times 120\right)}{\pi} = 108 \text{ V}$

2. The average voltage of the source is zero because the source waveform is symmetrical across the time axis.

3. $V_{rms} = \dfrac{V_{max}}{\sqrt{2}} = 120 \text{ V}$

4. $I_{rms} = \dfrac{V_{rms}}{R} = \dfrac{120}{10} = 12 \text{ A}$

5. $P = \dfrac{V_{rms}^2}{R} = 1.44 \text{ kW}$

5.1.2 Voltage-Controlled Circuits

The SCR converter in Figure 5.4 allows us to control the voltage across the load. The circuit consists of an ac source of potential v_s, a load resistance R, two diodes, and two SCRs. The waveforms of the circuit are shown in Figure 5.5. When the voltage across an SCR is positively biased and is triggered, the SCR is closed. When the current through the closed SCR falls below its holding value (near zero current), the SCR is opened. Because the source is a sinusoidal waveform, the current of the circuit is zero at π and 2π causing the corresponding SCR to open. The current flow in the circuit is similar to that explained for the full-wave rectifier circuit except that the current starts flowing after the triggering pulse is applied on S_1 at α during the positive half of the ac cycle, and on S_2 at $\alpha + 180°$ during the negative half of the cycle. The period during which the current flows through the SCR is called the "conduction period," γ.

$$\gamma = \pi - \alpha \tag{5.6}$$

The average voltage across the load is

$$V_{ave} = \frac{1}{2\pi}\int_0^{2\pi} v_t \, d\omega t = \frac{1}{\pi}\int_\alpha^\pi v_s \, d\omega t = \frac{V_{max}}{\pi}\int_\alpha^\pi \sin\omega t \, d\omega t = \frac{V_{max}}{\pi}(1+\cos\alpha) \tag{5.7}$$

The rms voltage across the load can be computed as

$$V_{rms} = \sqrt{\frac{1}{2\pi}\int_0^{2\pi} v_t^2 \, d\omega t} = \sqrt{\frac{V_{max}^2}{\pi}\int_\alpha^\pi \sin^2\omega t \, d\omega t} = \frac{V_{max}}{\sqrt{2}}\sqrt{\left(1 - \frac{\alpha}{\pi} + \frac{\sin 2\alpha}{2\pi}\right)} \tag{5.8}$$

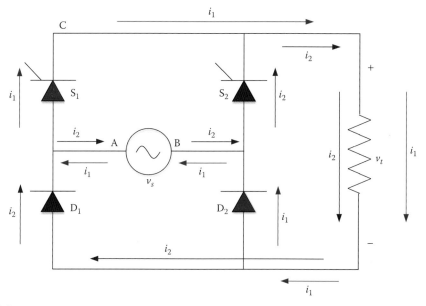

FIGURE 5.4
Full-wave SCR circuit.

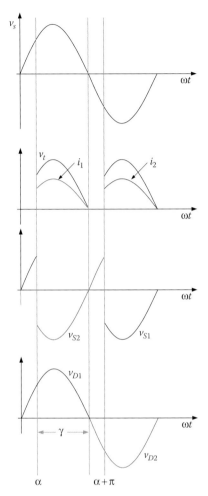

FIGURE 5.5
Waveforms of full-wave SCR circuit.

The rms voltage across the load is controlled from zero to $V_{max}/\sqrt{2}$ by adjusting the triggering angle α. The power across the load can be computed as

$$P = \frac{V_{rms}^2}{R} = \frac{V_{max}^2}{4\pi R}(2\gamma + \sin 2\alpha) \tag{5.9}$$

The voltages across the SCRs and diodes can be obtained from the loop of the source, where

$$v_s = v_{S1} + v_{S2} \tag{5.10}$$

Also,

$$v_s = v_{D1} + v_{D2} \tag{5.11}$$

Based on Equations 5.10 and 5.11, we can obtain the voltage waveforms across the SCRs and diodes, as shown in Figure 5.5. When one of the SCRs is conducting, the voltage across it is zero whereas the voltage across the other SCR is the source voltage. For the diodes, whether the SCRs are conducting or not, the voltage across one diode is the source voltage and the voltage across the other diode is zero.

EXAMPLE 5.2

A full-wave SCR converter circuit is used to regulate power across a 10 Ω resistance. The voltage source is 120 V (rms). What is the triggering angle that causes the load to consume 1 kW? Compute the average and rms currents of the load.

Solution:
The power expression in Equation 5.9 is nonlinear with respect to the triggering angle α.

$$P = \frac{V_{max}^2}{4\pi R}(2\gamma + \sin 2\alpha) = \frac{\left(\sqrt{2}\times 120\right)^2}{4\pi \times 10}\left[2(\pi - \alpha) + \sin 2\alpha\right] = 1000 \text{ W}$$

The solution of the above nonlinear equation is iterative.

$$\alpha = 71.9°$$

To compute the average current of the load, we need to compute the average voltage across the load as given by Equation 5.7.

$$V_{ave} = \frac{V_{max}}{\pi}(1 + \cos \alpha) = \frac{\sqrt{2}\times 120}{\pi}(1 + \cos 71.9°) = 70.8 \text{ V}$$

$$I_{ave} = \frac{V_{ave}}{R} = 7.08 \text{ A}$$

For the rms current, we can compute the rms voltage first, and then divide the voltage by the resistance of the load. Another simpler method is to use the power formula in Equation 5.5.

$$I_{rms} = \sqrt{\frac{P}{R}} = 10 \text{ A}$$

5.1.3 Three-Phase Circuits

Heavy loads are powered by three-phase, full-wave circuit as shown in Figure 5.6. The circuit consists of three switching legs, each has two SCRs. The center point of each leg is connected to one terminal of the three-phase source. The load is connected between the cathodes of the upper SCRs and the anodes of the lower SCRs. When one of the line-to-line voltages is greater than the other two, the corresponding SCRs can be triggered. For example, when the voltage v_{ab} is higher than v_{bc}, v_{ca}, v_{ba}, v_{cb}, and v_{ac}, we close S_1 and S_6, and the load voltage v_l is equal to v_{ab}. When the same logic is repeated for all other line-to-line voltages, we get the waveforms shown in Figure 5.7.

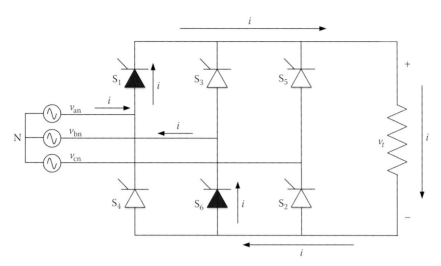

FIGURE 5.6
Three-phase full-wave ac/dc SCR switching circuit operating between points 1 and 2.

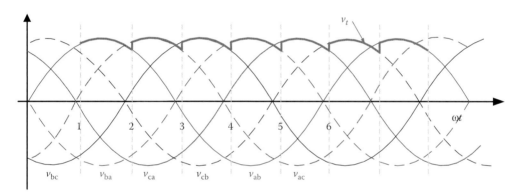

FIGURE 5.7
Three-phase full-wave ac/dc switching circuit waveforms.

The circuit has six switching intervals for each cycle; each one has a conduction period of 60°. Hence, the average voltage of the circuit

$$V_{ave} = 6\left(\frac{1}{2\pi}\int_{\alpha_{ab}}^{\alpha_{ab}+60°} v_{ab}\,d\omega t\right) = 6\left(\frac{1}{2\pi}\int_{\alpha_{ab}}^{\alpha_{ab}+60°} \sqrt{3}\,V_{max}\sin\omega t\,d\omega t\right) \quad (5.12)$$

$$V_{ave} = \frac{3\sqrt{3}\,V_{max}}{\pi}\sin(\alpha_{ab} + 30°)$$

where:
v_{ab} is the line-to-line voltage, which is the reference waveform
α_{ab} is the triggering angle measured from the zero crossing of the line-to-line voltage v_{ab}
V_{max} is the maximum value of the phase voltage (not line-to-line voltage)

Keep in mind that α_{ab} must be between 60° and 120°. This is the range at which v_{ab} is larger than all other voltages.

$$120° \geq \alpha_{ab} \geq 60° \tag{5.13}$$

The rms voltage across the load is

$$V_{rms} = \sqrt{\frac{6}{2\pi} \int_{\alpha_{ab}}^{\alpha_{ab}+60°} \left(\sqrt{3}\, V_{max} \sin \omega t\right)^2 d\omega t} = \frac{3V_{max}}{\sqrt{2}} \sqrt{\frac{1}{3} - \frac{\sqrt{3}}{2\pi} \cos\left(2\alpha_{ab} + 60°\right)} \tag{5.14}$$

EXAMPLE 5.3

For the circuit in Figure 5.6, assume the SCRs are replaced by diodes. If the line-to-line voltage of the circuit is 415.7 V and the load resistance is 10 Ω. Compute the average voltage across the load and the rms current of the load.

Solution:
For the diode circuit, the conduction period is 60° and α_{ab} is also 60°. Using Equation 5.12, we can compute the average voltage across the load as

$$V_{ave} = \frac{3\sqrt{3}\, V_{max}}{\pi} \sin(\alpha_{ab} + 30°) = \frac{3\sqrt{3}\left[\sqrt{2}\left(415.7/\sqrt{3}\right)\right]}{\pi} \sin(60 + 30°) = 561.39\ \text{V}$$

To compute the rms current, use Equation 5.14

$$V_{rms} = \frac{3V_{max}}{\sqrt{2}} \sqrt{\frac{1}{3} - \frac{\sqrt{3}}{2\pi} \cos\left(2\alpha_{ab} + 60°\right)}$$

$$= \frac{3\left[\sqrt{2}\left(415.7/\sqrt{3}\right)\right]}{\sqrt{2}} \sqrt{\frac{1}{3} - \frac{\sqrt{3}}{2\pi} \cos\left(120 + 60°\right)}$$

$$= 561.88\ \text{V}$$

Note that the rms voltage is almost the same as the average voltage. Can you tell why? The rms current is

$$I_{rms} = \frac{V_{rms}}{R_L} = \frac{561.88}{10} = 56.188\ \text{A}$$

The circuit in Figure 5.6 uses SCR as switching devices. Hence, the circuit controls only the triggering time as the commutation is determined by the fall of current below the holding value of the switch. If both the triggering angle (α) and the commutation angle (β) need to be controlled, we can use transistors instead of SCRs, as shown in Figure 5.8. The waveforms of the circuit are shown in Figure 5.9. In this circuit, the average voltage is

$$V_{ave} = 6\left(\frac{1}{2\pi} \int_{\alpha_{ab}}^{\beta_{ab}} v_{ab}\, d\omega t\right) = 6\left(\frac{1}{2\pi} \int_{\alpha_{ab}}^{\beta_{ab}} \sqrt{3}\, V_{max} \sin \omega t\, d\omega t\right) \tag{5.15}$$

$$V_{ave} = \frac{3\sqrt{3}\, V_{max}}{\pi} (\cos \alpha_{ab} - \cos \beta_{ab})$$

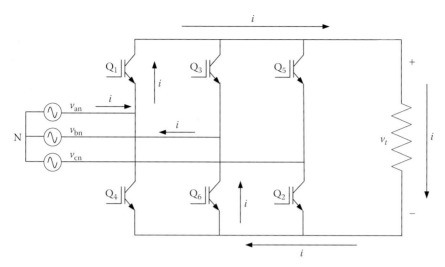

FIGURE 5.8
Three-phase full-wave ac/dc transistor switching circuit.

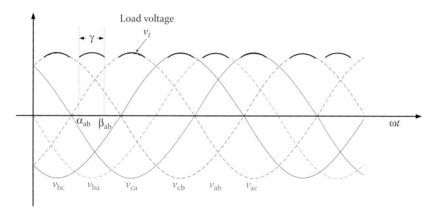

FIGURE 5.9
Waveforms of the circuit in Figure 5.8.

where:

v_{ab} is the line-to-line voltage of the source and is the reference waveform
α_{ab} is the triggering angle measured from the zero crossing of the line-to-line voltage v_{ab}
β_{ab} is the commutation angle measured from the zero crossing of the line-to-line voltage v_{ab}
V_{max} is the maximum value of the phase voltage (not line-to-line voltage)

The rms voltage across the load is

$$V_{rms} = \sqrt{\frac{6}{2\pi} \int_{\alpha_{ab}}^{\beta_{ab}} \left(\sqrt{3}\, V_{max} \sin \omega t\right)^2 d\omega t} = \frac{3 V_{max}}{2\sqrt{\pi}} \sqrt{2\gamma + \sin 2\alpha_{ab} - \sin 2\beta_{ab}} \qquad (5.16)$$

Keep in mind that α_{ab} and β_{ab} must be between $60°$ and $120°$.

5.2 AC/DC Converters with Inductive Load

For a purely inductive load, the current (i_L) lags the voltage (v) by 90°, as shown in Figure 5.10. On the same figure, the instantaneous power (ρ) consumed by the inductor is shown as well. It is the multiplication of the instantaneous voltage by the instantaneous current. The power that creates work is the average value of the instantaneous power, that is,

$$P = \frac{1}{2\pi} \int_0^{2\pi} \rho \, d\omega t \qquad (5.17)$$

where:
  P is the power that creates work (real power)
  ρ is the instantaneous power

Hence, the real power consumed by pure inductors (or capacitors) is zero. Moreover, the instantaneous power consumed during the first quarter of the voltage cycle is returned back to the source during the second quarter cycle. The inductor, in this case, does not store any energy.

Assume that a mechanical switch is used to interrupt the current of an inductor. When the switchblades open while the inductor energy is not fully returned to the source, the voltage across the switch increases rapidly to create an arc between the blades. The arc keeps the current flowing to dissipate the remainder of the energy of the inductor.

Consider the solid-state switching circuit in Figure 5.11 and its waveforms in Figure 5.12. The circuit consists of an inductive load powered by ac source through an SCR. Assume the SCR is triggered at angle α. Because of the presence of the inductor, we expect the current to extend beyond π to β, which is known as the "commutation angle." The SCR is commutated only when the energy of the inductor is returned back to the source.

If we ignore the voltage drop across the SCR during conduction, the voltage across the load v_t is equal to the source voltage v_s. Hence,

$$v_t = v_t(u_\alpha - u_\beta) \qquad (5.18)$$

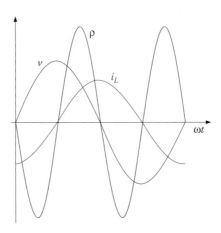

FIGURE 5.10
Instantaneous power of a purely inductive load.

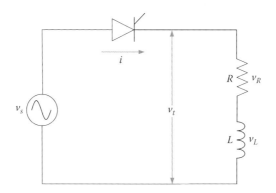

FIGURE 5.11
Switching inductive load.

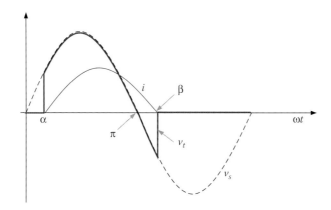

FIGURE 5.12
Waveforms of circuit in Figure 5.11.

where:

u_α is a unit step function; its value is zero unless $\omega t \geq \alpha$

u_β is a unit step function; its value is zero unless $\omega t \geq \beta$

5.2.1 Current Calculations

Because of the switching action, the current in the circuit is composed of various harmonics and can be computed using frequency domain analysis. The first step is to convert Equation 5.18 into frequency domain (*S* domain).

$$V_t(S) = \mathcal{L}v_t = \mathcal{L}v_t(u_\alpha - u_\beta)$$

$$V_t(S) = V_{max}\left[\frac{S\sin\alpha + \omega\cos\alpha}{S^2 + \omega^2}e^{-\alpha S/\omega} - \frac{S\sin\beta + \omega\cos\beta}{S^2 + \omega^2}e^{-\beta S/\omega}\right] \tag{5.19}$$

where:

$\mathcal{L}$ is the Laplace operator

The second step is to compute the load current in frequency domain.

$$I(S) = \frac{V_{max}}{R+SL}\left[\frac{S\sin\alpha + \omega\cos\alpha}{S^2 + \omega^2}e^{-\alpha S/\omega} - \frac{S\sin\beta + \omega\cos\beta}{S^2 + \omega^2}e^{-\beta S/\omega}\right] \tag{5.20}$$

The third step is to compute the current in time domain by Laplace inverse of Equation 5.20.

$$i(t) = \mathcal{L}^{-1}I(S) \tag{5.21}$$

During the conduction period, this current is

$$i(t) = \frac{V_{max}}{Z}\left[\sin(\omega t - \theta) + \sin(\theta - \alpha)e^{-(\omega t - \alpha)/\omega\tau}\right] \tag{5.22}$$

where:
  $\theta = \tan^{-1}(\omega\mathcal{L}/R)$ is the power factor angle of the load
  $\tau = L/R$ is the time constant of the load
  ω is the frequency of the supply voltage
  $Z = \sqrt{R^2 + (\omega L)^2}$ is the load impedance

The decaying component in Equation 5.22 is due to the presence of the inductance. The commutation angle is determined by setting the current at β to zero, hence

$$i(\beta) = \frac{V_{max}}{Z}\left[\sin(\beta - \theta) + \sin(\theta - \alpha)e^{-(\beta - \alpha)/\omega\tau}\right] = 0 \tag{5.23}$$

$$\sin(\beta - \theta) = \sin(\alpha - \theta)e^{-(\beta - \alpha)/\omega\tau}$$

The value of β can be found by the numerical solution of Equation 5.23
    The average current is

$$I_{ave} = \frac{1}{2\pi}\int_\alpha^\beta i(t)d\omega t = \frac{1}{2\pi}\int_\alpha^\beta \frac{V_{max}}{Z}\left[\sin(\omega t - \theta) + \sin(\theta - \alpha)e^{-(\omega t - \alpha)/\omega\tau}\right]d\omega t \tag{5.24}$$

$$I_{ave} = \frac{V_{max}}{2\pi Z}\left[\cos(\alpha - \theta) - \cos(\beta - \theta) - \omega\tau\sin(\theta - \alpha)\left(1 - e^{-\gamma/\omega\tau}\right)\right]$$

where:
  γ is the conduction period

$$\gamma = \beta - \alpha \tag{5.25}$$

The rms current is

$$I_{rms} = \sqrt{\frac{1}{2\pi}\int_\alpha^\beta i^2(t)d\omega t} = \frac{V_{max}}{2\pi Z}\sqrt{\int_\alpha^\beta\left[\sin(\omega t - \theta) + \sin(\theta - \alpha)e^{-(\omega t - \alpha)/\omega\tau}\right]^2 d\omega t} \tag{5.26}$$

The power consumed by the load is

$$P = R I_{rms}^2 = R \left(\frac{V_{max}}{2\pi Z} \right)^2 \int_{\alpha}^{\beta} \left[\sin(\omega t - \theta) + \sin(\theta - \alpha) e^{-(\omega t - \alpha)/\omega \tau} \right]^2 d\omega t \qquad (5.27)$$

5.2.2 Voltage Calculations

During the conduction period, from α to β, the current does not reverse its flow, as shown in Figure 5.12. Because the real power consumed by the inductor is zero, the voltage across the inductor v_L must reverse its polarity during the conduction period. Moreover, as shown in the case of a purely inductive load in Figure 5.10, while the current is not reversing, (for instance, between 90° and 270°), the average voltage across the inductor is zero.

$$V_{L\,ave} = \frac{1}{2\pi} \int_{90°}^{270°} v_s \, d\omega t = \frac{1}{2\pi} \int_{90°}^{270°} (V_{max} \sin \omega t) d\omega t = 0 \qquad (5.28)$$

The average voltage across the entire load is

$$V_{t\,ave} = V_{R\,ave} + V_{L\,ave} \qquad (5.29)$$

Because $V_{L\,ave} = 0$, the average voltage across the entire load is equal to the average voltage across the resistance only. Hence,

$$V_{t\,ave} = V_{R\,ave} = \frac{1}{2\pi} \int_{\alpha}^{\beta} (V_{max} \sin \omega t) d\omega t = \frac{V_{max}}{2\pi} (\cos \alpha - \cos \beta) \qquad (5.30)$$

The rms voltage across the load is

$$V_{t\,rms} = \sqrt{\frac{1}{2\pi} \int_{\alpha}^{\beta} v_s^2 \, d\omega t} = \sqrt{\frac{1}{2\pi} \int_{\alpha}^{\beta} (V_{max} \sin \omega t)^2 \, d\omega t}$$

$$(5.31)$$

$$V_{t\,rms} = \frac{V_{max}}{2} \sqrt{\frac{\gamma}{\pi} - \frac{\sin 2\beta - \sin 2\alpha}{2\pi}}$$

EXAMPLE 5.4

For the circuit in Figure 5.11, assume the SCRs is triggered at 60° and commutated at 120°. If the source voltage is 120 V and the load resistance is 10 Ω. Compute the average and rms voltages across the load.

Solution:
Using Equation 5.30, we can compute the average voltage across the load

$$V_{ave} = \frac{V_{max}}{2\pi} (\cos \alpha - \cos \beta) = \frac{\sqrt{2} \times 120}{2\pi} (\cos 60° - \cos 120°) = 27 \text{ V}$$

To compute the rms voltage across the load, we use Equation 5.31

$$V_{t\,rms} = \frac{V_{max}}{2} \sqrt{\frac{\gamma}{\pi} - \frac{\sin 2\beta - \sin 2\alpha}{2\pi}} = \frac{\sqrt{2} \times 120}{2} \sqrt{\frac{(\pi/3)}{\pi} - \frac{\sin(4\pi/3) - \sin(2\pi/3)}{2\pi}} = 66.22 \text{ V}$$

5.2.3 Freewheeling Diodes

Due to the presence of an inductance, the voltage across the load is negative between π and β, as shown in Figure 5.12. In some application, such as electric drives, this negative voltage produces opposing torque that is causing unnecessary extra load on the machine. Besides the extra losses, it may lead to vibrating speed and could overheat the machine. This is because the current continues to flow until the total energy of the inductor is zero. In Figure 5.12, the inductor acquires energy from π and until the current peaks. This is because di/dt is positive causing the voltage across the inductor $v_L = L(di/dt)$ to be positive. The inductor dissipates the acquired energy during the remainder of the conduction period because di/dt is negative causing the voltage of the inductor to reverse its polarity while the current is still flowing in the same direction.

To prevent the voltage across the load from becoming negative, a freewheeling diode, shown in Figure 5.13, is used as another path for the inductor to dissipate its energy instead of through the SCR. This way, the current through the SCR at π goes to zero and the SCR is commutated. The freewheeling diode is in parallel with the load and opposite in polarity to the SCR, as shown in Figure 5.13. To analyze the circuit, we divide the conduction period into two segments. The first one is from α to π, and the second one is from π to β. Figure 5.14 shows the circuits during these two segments. In the first segment, as shown in Figure 5.14a, the source voltage is positive and the current flow through the SCR to the load. Equation 5.22 describes the current during this segment. The voltage polarity of the load inductor is positive during the period from α until the current peaks. During this period, the inductor acquires energy. From the angle at the peak current to π, the voltage of the inductor reverses its polarity to return some of its energy back to the source. After π, the source voltage becomes negative, the SCR is opened, and the diode is in forward bias and closed. The remaining energy of the inductor is dissipated in the resistance through the circuit, as shown in Figure 5.14b. Between π and β, the current of the freewheeling diode can be described by a first-order decaying equation

$$i_d = i(\pi)e^{-(\omega t - \pi)/\omega\tau} \tag{5.32}$$

$i(\pi)$ in Equation 5.32 is the current of the load at π. This can be computed using Equation 5.22

$$i(\pi) = \frac{V_{max}}{Z}\left[\sin(\pi - \theta) + \sin(\theta - \alpha)e^{-(\omega t - \alpha)/\omega\tau}\right] + i(\alpha) \tag{5.33}$$

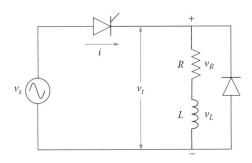

FIGURE 5.13
Circuit in Figure 5.11 with freewheeling diode.

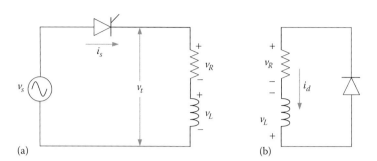

FIGURE 5.14
Active circuits with freewheeling diode: (a) from α to current peak angle and (b) from π to β.

where:

    $i(\alpha)$ is the initial current of the SCR circuit at α. In some cases, when the current is continuous, $i(\alpha)$ is not zero

The waveforms of the circuits in Figure 5.14 are shown in Figure 5.15. Note that the load voltage between π and β is zero as the diode is conducting. The commutation angle β may extend with the freewheeling diode. This can be seen from Equation 5.32.

$$\frac{i_d}{i(\pi)} = e^{-(\omega t - \pi)/\omega \tau} \tag{5.34}$$

Theoretically, this first-order equation decays at $\omega t = \infty$. Practically, we can assume that β is reached when the current is small enough, say 5% of the initial value. In this case, we can rewrite Equation 5.34 as

$$0.05 \approx e^{-(\beta - \pi)/\omega \tau} \tag{5.35}$$

Or

$$\beta \approx \pi - \omega \tau (\ln 0.05) \approx \pi + 3\omega \tau \tag{5.36}$$

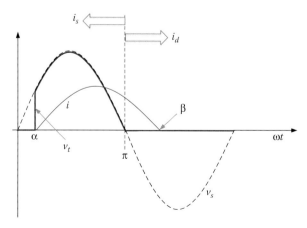

FIGURE 5.15
Waveforms of circuit with freewheeling diode.

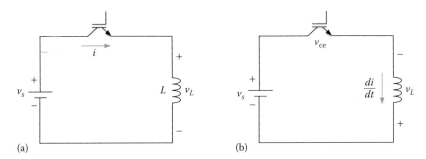

FIGURE 5.16
Switching modes of transistor with purely inductive load: (a) transistor is closed and (b) transistor is opening.

The freewheeling diode is also essential in dc switching circuits with inductive load. Without which the solid-state switch will be damaged. Take, for example, the circuits in Figure 5.16. Because of the dc source, when the transistor is triggered, that is, closed (see Figure 5.16a), the energy goes from the source to the inductor. If the transistor if forced to open (see Figure 5.16b), the energy of the inductor has no place to go. Hence, a sequence of events will occur to dissipate the energy of the inductor and will damage the transistor as follows:

- When the transistor is opened, the current initially falls quickly, hence di/dt is negative and very large.
- This forces the voltage across the inductor to reverse quickly and become very large as $v_L = L(di/dt)$.
- The voltage across the switch (collector to emitter, v_{ce}), as shown in Figure 5.16b, is

$$v_{ce} = v_s + v_L = v_s + L\frac{di}{dt} \tag{5.37}$$

- The transient voltage $L(di/dt)$ is negative and its magnitude is much larger than the source voltage v_s. Hence, v_{ce} is excessively large and is in the reverse breakdown mode across the transistor. This could cause the transistor to fail short to allow the energy of the inductor to return back to the source. Also, the rate of change of the voltage across the transistor could cause the damage as every solid-state device has a limit on its dv/dt.

To solve this problem, a freewheeling diode is used in anti-parallel with the transistor, as shown in Figure 5.17. With this arrangement, when the transistor is closed, the current i_c pass through the transistor to the load. When the transistor is opened, the current i_o, pass through the diode and return the inductor energy back to the source.

5.3 DC/DC Converters

The dc/dc converters provide adjustable dc voltage waveforms at the output. These converters are widely used in types 3 and 4 wind turbine systems. There are three basic dc/dc converters:

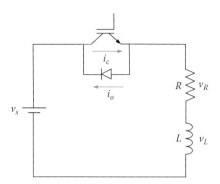

FIGURE 5.17
Transistor switching in dc circuit with freewheeling diode.

- *Buck converter*: This is a step-down converter where the output voltage is less than the input voltage.
- *Boost converter*: This is a step-up converter where the output voltage is higher than the input voltage.
- *Buck–boost converter*: This is a step-down/step-up converter where the output voltage can be made either lower or higher than the input voltage.

5.3.1 Buck Converter

The buck converter is also known as "chopper" and is widely used in the dc link of doubly fed induction generator systems. Figure 5.18 shows a simple chopper circuit consisting of dc power source (V_{dc}), a load, and a transistor. When the transistor is closed, the voltage across the load is equal to the source voltage. When the transistor is open, no current flows in the circuit and the load voltage is zero, as shown in the waveform in Figure 5.18. The transistor is closed for the time t_{on}, and open for t_{off}. The period of switching is τ. The average voltage across the load V_{ave} is

$$V_{ave} = \frac{1}{\tau} \int_0^{t_{on}} V_{dc}\, dt = \frac{t_{on}}{\tau} V_{dc} = K V_{dc} \tag{5.38}$$

where:
  $K = t_{on}/\tau$ is the *duty ratio* of the switching cycle

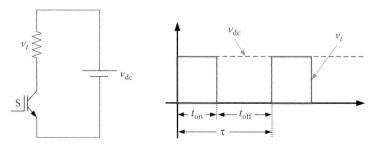

FIGURE 5.18
Simple chopper circuit.

EXAMPLE 5.5

The switching frequency of a chopper is 10 KHz, and the source voltage is 40 V. For an average load voltage of 20 V and a load resistance of 10 Ω, compute the following:

1. Duty ratio
2. On-time and the switching period
3. Average voltage across the transistor
4. Average current of the load
5. rms voltage across the load
6. Load power

Solution:

1. As given in Equation 5.38, the duty ratio is

$$K = \frac{V_{\text{ave}}}{V_s} = \frac{20}{40} = 0.5$$

2. Before we compute the on-time, we need to compute the period.

$$\tau = \frac{1}{f} = \frac{1}{10} = 0.1 \, \text{ms}$$

$$t_{\text{on}} = K^\tau = 0.5 * 0.1 = 0.05 \, \text{ms}$$

3. The average voltage across the transistor $V_{\text{ave-tr}}$ is

$$V_{\text{ave-tr}} = V_s - V_{\text{ave}} = 40 - 20 = 20 \, \text{V}$$

4. The average load current is

$$I_{\text{ave}} = \frac{V_{\text{ave}}}{R} = \frac{20}{10} = 2 \, \text{A}$$

5. The rms voltage of the load can be computed using the formula for the rms quantity

$$V = \sqrt{\frac{1}{\tau} \int_0^{t_{\text{on}}} V_s^2 dt} = \sqrt{\frac{V_s^2}{\tau} t_{\text{on}}} = V_s \sqrt{\frac{t_{\text{on}}}{\tau}} = 28.28 \, \text{V}$$

6. $P = \dfrac{V^2}{R} = \dfrac{28.28^2}{10} = 80 \, \text{W}$

5.3.2 Boost Converter

With the boost converter, the output voltage is higher than the input voltage. A simple circuit for the boost converter is shown in Figure 5.19a. It consists of a dc voltage source, inductor, switching transistor, diode, capacitor, and load resistance. The transistor is switched at high frequency (several kHz). It is closed for a time interval t_{on} and is open

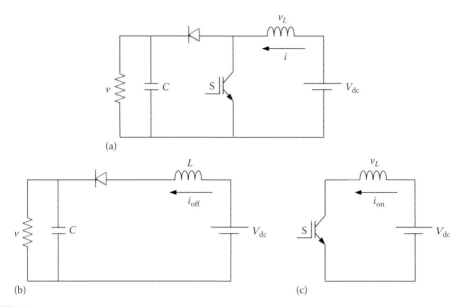

FIGURE 5.19
(a–c) A simple boost converter.

during t_{off}. When the transistor is closed, the current i_{on} charges the inductor, as depicted in Figure 5.19c. In this process, the voltage across the inductor v_L is

$$v_L = L\frac{di_{on}}{dt} \approx L\frac{\Delta i_{on}}{t_{on}} \tag{5.39}$$

where:
Δi_{on} is the change in the current during the period t_{on}

Because the voltage across the inductor is equal to the source voltage when the transistor is closed, Equation 5.39 can be rewritten as

$$V_{dc} = L\frac{\Delta i_{on}}{t_{on}} \tag{5.40}$$

When the transistor is opened, the load energy comes from two sources: the dc source and the inductor. The stored energy in the inductor while the transistor is closed is transferred to the load through the diode. The inductor current in this case is i_{off}, as depicted in the circuit in Figure 5.19b. The load voltage in this case is

$$v = V_{dc} - v_L = V_{dc} + L\frac{\Delta i_{off}}{t_{off}} \tag{5.41}$$

Because the inductor is producing energy during t_{off}, and because the current through the inductor is not reversed, the voltage polarity across the inductance must reverse. This is the reason for the positive sign in front of L in Equation 5.41. During steady state,

$$\Delta i_{on} = \Delta i_{off} = \Delta i \tag{5.42}$$

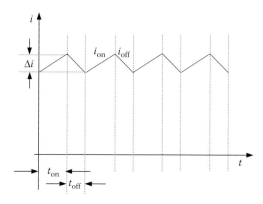

FIGURE 5.20
Waveform of boost converter.

Figure 5.20 shows the waveform of the current during steady state. Equation 5.41 can be rewritten as

$$v = V_{dc} + L \frac{\Delta i_{off}}{t_{off}} = V_{dc} + L \frac{\Delta i_{on}}{t_{off}}$$ (5.43)

Substituting the value of Δi_{on} in Equation 5.40 into the above equation yields

$$v = V_{dc} \left(1 + \frac{t_{on}}{t_{off}} \right)$$ (5.44)

Equation 5.44 shows that the voltage across the load can be controlled by adjusting t_{on} and t_{off}. Furthermore, the load voltage is always higher than the source voltage unless $t_{on} = 0$.

The capacitor in the circuit is used as a filter to reduce the voltage ripples across the load. The larger the capacitance, the lesser the ripples. The diode is used to block the capacitor from discharging through the transistor when it is closed. Hence, the voltage in Equation 5.44 can be assumed to be the average value. The input and output powers are

$$P_{in} = V_{dc} I$$
$$P_{out} = V I_{off}$$ (5.45)

where:
    P_{in} is the input power from the dc source
    P_{out} is the output power consumed by the load
    I is the average currents of the source
    I_{off} is the average currents of the load when the transistor is open
    V is the average voltage across the load

If we assume that the components of the circuit are ideal, the input power to the converter is equal to the output power. Hence,

$$V_{dc} I = V I_{off}$$ (5.46)

EXAMPLE 5.6

A boost converter is used to step up 20 V source to 50 V across a load. The switching frequency of the transistor is 5 kHz, and the load resistance is 10 Ω. Compute the following:

1. Value of the inductance that would limit the current ripple at the source side to 100 mA
2. Average current of the load
3. Power delivered by the source
4. Average current of the source

Solution:

1. We can use Equation 5.40 to compute the inductance. But first, we need to compute t_{on} using Equation 5.44

$$v = V_{dc}\left(1 + \frac{t_{on}}{t_{off}}\right)$$

$$50 = 20\left(1 + \frac{t_{on}}{t_{off}}\right)$$

$$t_{on} = 1.5 t_{off}.$$

Because the switching frequency is 5 kHz

$$\tau = \frac{1}{f} = \frac{1}{5} = 0.2\,\text{ms}$$

then,

$$t_{on} = 1.5 t_{off} = 1.5 \times (0.2 - t_{on})$$

$$t_{on} = 0.12\,\text{ms}$$

To compute the value of the inductance, we can use Equation 5.40.

$$V_{dc} = L\frac{\Delta i_{on}}{t_{on}}$$

$$20 = L\frac{100}{0.12}$$

$$L = 24\,\text{mH}$$

2. $I_{off} = \dfrac{V}{R} = \dfrac{50}{10} = 5\,\text{A}$

3. The power delivered by the source is the same power consumed by the load assuming that the system components are all ideal.

$$P = VI_{off} = 50 \times 5 = 250\,\text{W}$$

4. The average current of the source can be computed using the power equation in the previous step

$$I = \frac{P}{V_{dc}} = \frac{250}{20} = 12.5\,\text{A}$$

5.3.3 Buck–Boost Converter

The buck–boost converter has the same components as the boost converter, but is structured differently, as shown in Figure 5.21a. When the transistor is closed, as shown in Figure 5.21c, the current i_{on} flows through the inductor, and energy is acquired by the inductor. When the transistor is opened, the inductor delivers its stored energy to the load, as shown in Figure 5.21b. The capacitor is used as a filter to minimize the voltage ripples across the load, and the diode is used to prevent the load from being energized when the transistor is closed. The capacitor current can be assumed very small compared with the load current.

When the transistor is closed, the voltage across the inductor is

$$v_L = L \frac{di_{on}}{dt} \approx L \frac{\Delta i_{on}}{t_{on}} \tag{5.47}$$

When the transistor is closed, the voltage across the inductor is equal to the source voltage. Hence,

$$V_{dc} = L \frac{\Delta i_{on}}{t_{on}} \tag{5.48}$$

When the transistor is open, the inductor dissipates its acquired energy into the load. Because the current through the inductor is not reversed, the voltage across the inductor must reverse its polarities to discharging its energy. During this period, the voltage across the inductor is

$$v_L = -L \frac{\Delta i_{off}}{t_{off}} \tag{5.49}$$

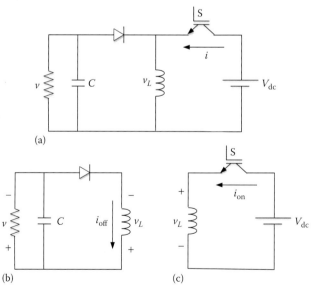

FIGURE 5.21
(a–c) A simple buck–boost converter.

The voltage across the load when the transistor is open is equal to the inductor voltage. Hence,

$$v = -L \frac{\Delta i_{\text{off}}}{t_{\text{off}}} \tag{5.50}$$

At steady state when $\Delta i_{\text{on}} = \Delta i_{\text{off}}$, Equations 5.48 and 5.50 can be combined as

$$v = -V_{\text{dc}} \frac{t_{\text{on}}}{t_{\text{off}}} \tag{5.51}$$

As seen in Equation 5.51, the load voltage can be controlled by adjusting the ratio of $t_{\text{on}}/t_{\text{off}}$. If the on-time is zero, the load voltage is zero. If $0 < t_{\text{on}} < t_{\text{off}}$, the load voltage is lower than the source voltage. If $t_{\text{on}} = t_{\text{off}}$, the load voltage is equal to the source voltage. If $t_{\text{on}} > t_{\text{off}}$, the load voltage is greater than the source voltage. Keep in mind that Equation 5.51 is not valid for $t_{\text{off}} = 0$ when the transistor is continuously closed. The system in this case is unstable as the current i_{on} reaches very high values because the inductor is short circuiting the source and cannot acquire any more energy.

If we assume that the circuit's components are ideal, the input power to the converter is equal to the output power. Hence,

$$P_{\text{in}} = P_{\text{out}} = V_{\text{dc}} I_{\text{on}} = V I_{\text{off}} \tag{5.52}$$

EXAMPLE 5.7

A buck–boost converter with an input voltage of 20 V is used to regulate the voltage across a 10 Ω load. The switching frequency of the transistor is 5 kHz. Compute the following:

1. On-time of the transistor to maintain the output voltage at 10 V
2. Output power
3. On-time of the transistor to maintain the output voltage at 40 V
4. Output power

Solution:

1. The period can be computed from the switching frequency f

$$\tau = \frac{1}{f} = \frac{1}{5} = 0.2 \text{ ms}$$

Use Equation 5.51 to compute the time ratio. We only need to use the magnitudes of the voltages.

$$\frac{t_{\text{on}}}{t_{\text{off}}} = \frac{V}{V_{\text{dc}}} = \frac{10}{20} = 0.5$$

Hence,

$$t_{\text{on}} = 0.0667 \text{ ms}$$

2. The output power

$$P = \frac{V^2}{R} = 10 \text{ W}$$

3. $\dfrac{t_{on}}{t_{off}} = \dfrac{V}{V_{dc}} = \dfrac{40}{20} = 2$

Hence,

$$t_{on} = 0.1333 \text{ ms}$$

4. The output power

$$P = \dfrac{V^2}{R} = \dfrac{40^2}{10} = 160\,\text{W}$$

5.4 DC/AC Converters

The dc/ac converter (also known as an "inverter") is one of the main components of types 3 and 4 wind turbines. The conversion from dc to ac is done by switching transistors in a specific pattern to create adjustable voltage waveforms in terms of magnitude, frequency, sequence, and phase shift. To reduce the voltage ripples and harmonics, the transistors are switched at high frequency and their duty ratios are continuously adjusted over a single ac period. For most wind applications, the system is three-phase, which is the one that is covered in this chapter.

5.4.1 Three-Phase DC/AC Converter

Three-phase waveforms can be obtained using the converter shown in Figure 5.22. It is composed of six transistors, a dc source, and a three-phase load. The circuit is also known as "six-pulse inverter." The transistors are switched according to the sequence at the top part of Figure 5.23. Each cycle is divided into six intervals; each interval time t_i corresponds to $60°$.

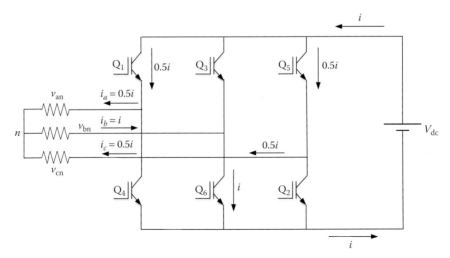

FIGURE 5.22
Three-phase dc/ac converter.

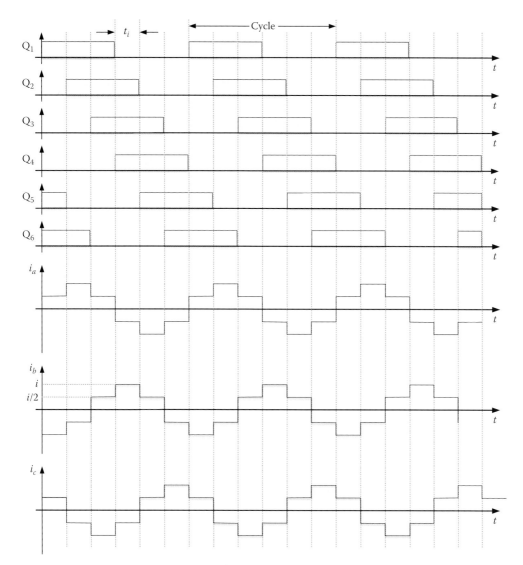

FIGURE 5.23
Timing of transistors and the phase currents.

Each transistor is closed for three consecutive intervals (180°) and opened for the following three intervals. The switching of the transistors is based on their ascending numbers. For example, transistor Q_1 is closed first and then Q_2 is closed after one time interval, Q_3 is closed after additional time interval, and so on.

Figure 5.22 shows the flow of current through the first time interval where transistors Q_1, Q_5, and Q_6 are closed. During this interval, the current from the source i is divided into two equal parts; one passes through Q_1 and the other through Q_5. At the neutral point n, the two currents are added and returned to the source through Q_6. The load currents during this interval are

$$i_a = i_c = \frac{i}{2}$$

$$i_b = -i$$

(5.53)

If you follow the same procedure for the other intervals, you will get the current wave-forms in Figure 5.23. These waveforms represent a balanced three-phase system.

The frequency of the load current is dependent on the interval time t_i. Because the cycle is composed of six intervals, the frequency of the ac waveform is

$$f = \frac{1}{6t_i} \tag{5.54}$$

To change the frequency of the ac waveform, the time of the interval must change.

EXAMPLE 5.8

A three-phase dc/ac converter is used to power a three-phase, Y-connected, resistive load of 10 Ω (in each phase). The dc voltage is 300 V. Compute the following:

1. Time interval if the desired frequency of the ac waveform is 200 Hz
2. Current of the dc source
3. Current of the load in rms
4. Voltage across each phase of the load in rms

Solution:

1. Use Equation 5.54 to compute the interval t_i.

$$t_i = \frac{1}{6f} = \frac{1}{6 \times 200} = 833 \ \mu s$$

2. During any given interval, the load has two of its phases in parallel, and the combination is in series with the third phase. For the interval in Figure 5.22, the load resistances of phases a and c are in parallel, and this combination is in series with the resistance of phase b. Similar combinations are true during all other time intervals. Hence,

$$R_{\text{total}} = 10 + 5 = 15 \ \Omega$$

Then the source current is

$$i = \frac{V_{\text{dc}}}{R_{\text{total}}} = \frac{300}{15} = 20 \text{ A}$$

3. The rms current of phase a is

$$I_a = \sqrt{\frac{1}{6t_i} \int_0^{6t_i} i_a^2 \, dt}$$

If you examine the waveform of i_a for one period, you will discover that the magnitude of the current is equal to i during two time intervals, and is equal to $i/2$ during four intervals. Hence, the rms current of phase a can be written as

$$I_a = \sqrt{\frac{1}{6t_i}\left[\int_0^{2t_i} i^2 dt + \int_0^{4t_i}\left(\frac{i}{2}\right)^2 dt\right]} = \frac{i}{\sqrt{2}} = 14.14 \text{ A}$$

This rms current is the same as that computed for the purely sinusoidal waveform where i is the maximum current of the instantaneous waveform.

4. The rms phase voltage of the load is

$$V_{an} = I_a R = 141.1 \text{ V}$$

5.4.2 Pulse Width Modulation

With the pulse width modulation (PWM) technique, the transistors are switched in a specific pattern to control the following four variables:

- Magnitude of output voltage
- Frequency of output voltage
- Phase sequence of output voltages
- Phase shift of output voltage

Before we discuss the PWM method, let us examine the switching of one leg of a three-phase dc/ac converter shown in Figure 5.24. The switching of Q_1 and Q_4 determines the voltage v_{ao} according to the following logic:

$$\text{when } Q_1 \text{ is closed and } Q_4 \text{ is opened}, v_{ao} = V_{dc}$$

$$\text{when } Q_4 \text{ is closed and } Q_1 \text{ is opened}, v_{ao} = 0 \qquad (5.55)$$

Figure 5.25 shows a one switching cycle of these two transistors. During this cycle the average voltage of the cycle is

$$v_{ao} = m V_{dc}$$

$$m = \frac{t_{on}}{\tau} = t_{on} f \qquad (5.56)$$

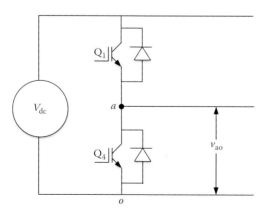

FIGURE 5.24
Leg of a three-phase ac/dc converter.

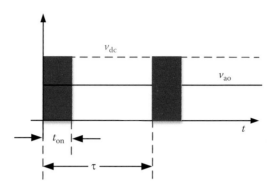

FIGURE 5.25
Switching of transistors in Figure 5.24.

where:
t_{on} is the on-time of Q_1
τ is the switching period
f is the switching frequency
m is the modulation index, or duty ratio

The PWM circuit generates four low-voltage signals: one high-frequency carrier and three reference signals. The reference signals are balanced three-phase waveforms with frequency equal to the desired frequency of the output voltage.

$$v_{ar} = V_{max-r} \sin\left(2\pi f_s t\right)$$

$$v_{br} = V_{max-r} \sin\left(2\pi f_s t - \frac{2\pi}{3}\right)$$

(5.57)

$$v_{cr} = V_{max-r} \sin\left(2\pi f_s t + \frac{2\pi}{3}\right)$$

where:
v_{ar} is the reference signal of phase a
V_{max-r} is the peak value of the reference signals
f_s is the frequency of the reference signals

Figure 5.26 shows the waveforms of the carrier and two reference signals for phases a and b. The carrier is fixed voltage and fixed frequency waveform. However, the voltage and frequency of the reference signals are adjustable. The PWM circuit determines the switching status of the transistors in Figure 5.22. For each phase, when the voltage of the carrier is less than the voltage of its reference signal, the upper transistor in the leg of that phase is closed and the lower one is opened. When the carrier voltage is higher than the reference voltage, the upper transistor is opened and the lower transistor is closed. Doing that for the complete cycle of the reference signal leads to the voltage waveforms v_{ao} and v_{bo} in Figure 5.26.

Subtracting v_{bo} from v_{ao} leads to the line to line voltage v_{ab}.

$$v_{ab} = v_{ao} - v_{bo}$$

(5.58)

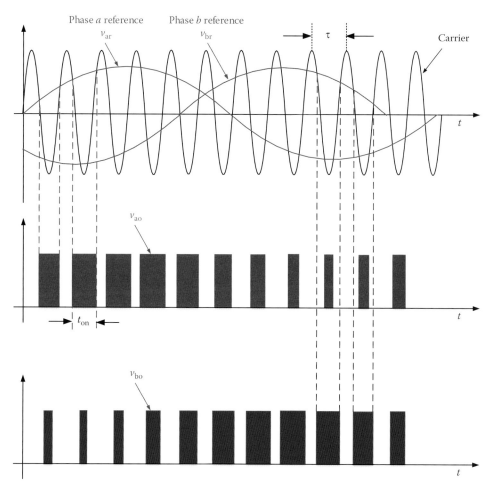

FIGURE 5.26
PWM signals and output voltages.

The waveform of v_{ab} is shown in Figure 5.27. It consists of pulses with different widths. If we compute the harmonics of these pulses, we get a large fundamental component and several higher frequency harmonics that can be represented by

$$v_{ab} = v_{ao} - v_{bo} = k V_{dc} \sin(2\pi f_s t) + \text{harmonic terms} \qquad (5.59)$$

where:
  k is the ratio of the peak values of the reference to carrier signals

$$k = \frac{V_{max-r}}{V_{max-car}}$$

If we ignore the harmonics, the fundamental components of all phases are balanced three-phase waveforms

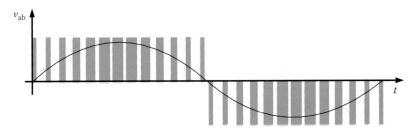

FIGURE 5.27
Line-to-line voltage with PWM.

$$v_{ab} = v_{ao} - v_{bo} \approx k V_{dc} \sin(2\pi f_s t)$$

$$v_{bc} = v_{bo} - v_{co} \approx k V_{dc} \sin\left(2\pi f_s t - \frac{2\pi}{3}\right) \qquad (5.60)$$

$$v_{ca} = v_{co} - v_{ao} \approx k V_{dc} \sin\left(2\pi f_s t + \frac{2\pi}{3}\right)$$

For some high-performance applications, the harmonics in Equation 5.59 could be higher than desired. To reduce the harmonics, more elaborate circuits must be used. One of them uses a very high-frequency carrier and modifies the conduction period in each cycle of the carrier signal. To explain this, let us consider v_{ao} in Equation 5.56 to be the instantaneous value which is the average value for one cycle of the high-frequency carrier signal. This is reasonable approximation as the reference signal is near the power line frequency and the carrier signal is often higher than 5 kHz.

In addition to the high frequency carrier, we assign separate modulation indices (duty ratio) for the three phases.

$$v_{ao} = m_a V_{dc}$$

$$v_{bo} = m_b V_{dc} \qquad (5.61)$$

$$v_{co} = m_c V_{dc}$$

$$m_a = \frac{t_{on-a}}{\tau} = t_{on-a} f$$

$$m_b = \frac{t_{on-b}}{\tau} = t_{on-b} f \qquad (5.62)$$

$$m_c = \frac{t_{on-c}}{\tau} = t_{on-c} f$$

where:
  t_{on-a} is the on-time of phase *a* during one period of the carrier signal
  τ is the period of the carrier signal
  f is the frequency of the carrier signal

Next, let us make the modulation index time dependent and in the form of

$$m_a = 0.5 + k \sin(2\pi f_s t)$$

$$m_b = 0.5 + k\sin\left(2\pi f_s t - \frac{2\pi}{3}\right)$$

$$m_c = 0.5 + k\sin\left(2\pi f_s t + \frac{2\pi}{3}\right) \qquad (5.63)$$

$$k = \frac{V_{\max-r}}{V_{\max-car}} = \frac{V_r}{V_{car}}$$

where:

k is the gain ratio: the ratio of the peak (or rms) values of the reference to carrier signals
V_r is the reference voltage in rms
V_{car} is the carrier voltage in rms
f_s is the frequency of the reference signals

In this case, the line to line voltages are

$$v_{ab} = v_{ao} - v_{bo} = m_a V_{dc} - m_b V_{dc} = \sqrt{3}\, k\, V_{dc}\sin\left(2\pi f_s t + \frac{\pi}{6}\right)$$

$$v_{bc} = v_{bo} - v_{co} = m_b V_{dc} - m_c V_{dc} = \sqrt{3}\, k\, V_{dc}\sin\left(2\pi f_s t - \frac{\pi}{2}\right) \qquad (5.64)$$

$$v_{ca} = v_{co} - v_{ao} = m_c V_{dc} - m_a V_{dc} = \sqrt{3}\, k\, V_{dc}\sin\left(2\pi f_s t + \frac{5\pi}{6}\right)$$

The above equation shows a three-phase balanced voltages that are free from harmonics. The phase to neutral voltages at the output are

$$v_{an} = k V_{dc}\sin\left(2\pi f_s t\right)$$

$$v_{bn} = k V_{dc}\sin\left(2\pi f_s t - \frac{2\pi}{3}\right) \qquad (5.65)$$

$$v_{cn} = k V_{dc}\sin\left(2\pi f_s t + \frac{2\pi}{3}\right)$$

In rms quantities

$$\bar{V}_{an} = \frac{k V_{dc}}{\sqrt{2}}\,\angle 0°$$

$$\bar{V}_{bn} = \frac{k V_{dc}}{\sqrt{2}}\,\angle -120° \qquad (5.66)$$

$$\bar{V}_{cn} = \frac{k V_{dc}}{\sqrt{2}}\,\angle 120°$$

Based on the above equation, we can control the voltage, frequency, phase shift, and phase sequence as follows:

- Voltage control is implemented by adjusting k
- Frequency control is implemented by adjusting the frequency of the reference signal f_s

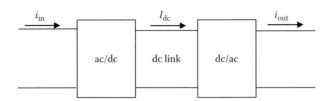

FIGURE 5.28
The ac/ac converter with dc link.

- Phase shift is implemented by shifting the reference signals
- Phase sequence is implemented by swapping the switching signals of any two legs in Figure 5.22

This powerful PWM technique is widely used in dc/ac applications such as adjustable speed drives, wind turbines, and power supplies.

5.5 AC/AC Converters

An ac/ac converter is shown in Figure 5.28. The system consists of two ac/dc converters connected back-to-back, as shown in the figure. In the first converter, the input ac waveform is converted into dc through the first ac/dc converter. The dc is then converted back to ac waveform through the dc/ac converter. The connection between the two converters is called "dc link." This is mainly the configuration used in types 3 and 4 wind turbines.

Exercise

1. A bipolar transistor is connected to a resistive load. The source voltage $V_{CC} = 40$ V and $R_L = 10$ Ω. In the saturation region, the collector–emitter voltage $V_{CE} = 1.0$ V and $\beta = 5$. While the transistor is in the saturation region, calculate the following:
   a. Load current
   b. Load power
   c. Losses in the collector circuit
   d. Losses in the base circuit
   e. Efficiency of the circuit

2. For the transistor in the previous problem, compute the load power and the efficiency of the circuit when the transistor is in the cutoff region. Assume that the collector current is 10 mA in the cutoff region.

3. A bipolar junction transistor operating in the saturation region has a base current of 10 A and a collector current of 50 A. Compute the following:
   a. Current gain of the transistor in the saturation region
   b. Losses of the transistor

4. Compute the rms voltage of the following waveform.

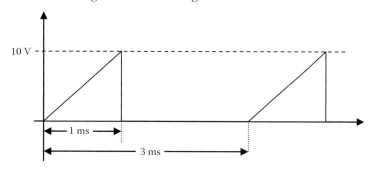

5. A half-wave rectifier circuit converts a 120 V (rms) into dc. The load of the circuit is 5 Ω resistance. Compute the following:
   a. Average voltage across the load
   b. Average voltage of the source
   c. rms voltage of the load
   d. rms current of the load
   e. Power consumed by the load

6. A half-wave SCR converter circuit is used to regulate the power across a 10 Ω resistance. When the triggering angle is 30°, the power consumed by the load is 500 W. Compute the rms voltage of the source, and the average and rms currents of the load.

7. An ac/dc half-wave SCR circuit is used to energize a resistive load. At a triggering angle of 30°, the average voltage across the load is 45 V.
   a. Compute the source voltage in rms
   b. If a full-wave circuit is used, while the triggering angle is maintained at 30°, compute the average voltage across the load

8. A full-wave SCR converter circuit is used to regulate the power across a 10 Ω resistance. The voltage source is 120 V (rms). At a triggering angle of 72°, compute the power across the load and the rms currents of the load.

9. A 120 V full-wave SCR battery charger is designed to provide 1.0 A of charging current. The battery has 1 Ω internal resistance. At the beginning of the charging process the voltage of the battery set is 60 V. Compute the triggering angle of the converter.

10. A full-wave, AC/DC converter is connected to a resistive load of 5 Ω. The voltage of the ac source is 120 V (rms). If the triggering angle of the converter is 90°, compute the rms voltage across the load and the power consumed by the load.

11. A boost converter has a source voltage $v_s = 20$ V and a load resistance $R = 10$ Ω. If the duty ratio of the transistor is 50%, compute the voltage across the load and the power consumed by the load.

12. A boost converter is used to step up 25 into 40 V. The switching frequency of the transistor is 1 kHz, and the load resistance is 100 Ω. Compute the following:
   a. Current ripple when the inductor is 30 mH
   b. Average current of the load
   c. Power delivered by the source

13. A buck–boost converter has an input voltage of 30 V. The switching frequency of its transistor is 5 kHz, and its duty ratio is 25%. Compute the average voltage across the load.

14. A buck–boost converter with an input voltage of 40 V is used to regulate the load voltage from 10 to 80 V. The on-time of the transistor is always fixed at 0.1 ms and the switching frequency is adjusted to regulate the load voltage. Compute the range of the switching frequency.

15. A three-phase dc/ac converter is used to power a three-phase, Y-connected, resistive load of 50 Ω (per phase). The dc voltage is 150 V. Compute the following:

    a. Frequency of the ac waveform if the time interval is 100 μs

    b. Current of the dc source

    c. Rms current of the load

    d. Rms of the line-to-line voltage across the load

16. A full-wave ac/dc four-SCR bridge converter circuit is used to power a resistive load of 10 Ω. The ac voltage is 120 V (rms), and the triggering angle of the SCR is adjusted to 60°. Calculate the following:

    a. Conduction period

    b. Average voltage across the load

    c. Average voltage across the SCRs

    d. Rms voltage across the load

    e. Average current

    f. Load power

17. A dc/dc converter consists of a 100 V dc source in series with 10 Ω load resistance and a bipolar transistor. For each cycle, the transistor is turned on for 200 μs and turned off for 800 μs. Calculate the following:

    a. Switching frequency of the converter

    b. Average voltage across the load

    c. Average load current

    d. Rms voltage across the load

    e. Rms current of the load

    f. Rms power consumed by the load

18. A dc/dc buck converter has an input voltage of 100 V and a duty ratio of 0.2. Compute the following:

    a. Load voltage

    b. Switching frequency of the converter if the on-time is 0.1 ms

19. An inductive load is connected to a full-wave four-SCR bridge circuit. The source is alternating current of 120 V. The resistive component of the load has an average voltage of 50 V.

    a. Compute the average voltage across any SCR

    b. Triggering circuit of one SCR failed and that SCR is not conducting anymore. If the triggering angle of the rest of the SCRs is unchanged, compute the average voltage of the load

20. A resistive load of 5 Ω is connected to an ac source of 120 V (rms) through a anti-parallel SCR circuit (two parallel SCRs with the anode of each one is connected to the cathode of the other). The triggering angle of the forward SCR (the one that conducts at the positive half of the cycle) is 30°. The triggering angle of the other SCR is (180° + 30°). Calculate the following:

    a. Average voltage across the load
    b. Power consumed by the load

21. A half-wave, single-phase ac to dc converter is power a load of 10 mH inductance in series with 10 Ω resistance. The ac voltage is 110 V (rms). For triggering angle of 30° calculate the following:

    a. Conduction period
    b. Average current of the load
    c. Average voltage of the load

22. For $\alpha = 10°$, repeat previous problem assuming that a freewheeling diode is used.

23. A single-phase, half-wave SCR circuit is used to control the power consumption of an inductive load. The resistive component of the load is 5 Ω. The source voltage is 120 V (rms). When the triggering angle is adjusted to 60°, the average current of the load is 6 A.

    a. Calculate the average voltage across the load
    b. Calculate the conduction period in degrees

24. A 120 V (rms), 60 Hz source is connected to a full-wave bridge. The load is represented by a resistance of 1 Ω in series with an inductive reactance of 3 Ω. At a triggering angle of 60°, the current of the load is continuous. Calculate the following:

    a. Average voltage across the load
    b. Average voltage across the resistive element of the load
    c. Average current of the load

25. An inductive load consists of a resistance and an inductive reactance connected in series. The circuit is excited by a full-wave ac/dc SCR converter. The ac voltage (input to the converter) is 120 V (rms), and the circuit resistance is 5 Ω. At a triggering angle of 30°, the load current is continuous. Calculate the following:

    a. Average voltage across the load
    b. Average load current
    c. rms voltage across the load

26. A purely inductive load of 10 Ω is connected to an ac source of 120 V (rms) through a half-wave SCR circuit.

    a. If the SCR is triggered at 90°, calculate the angle at the maximum instantaneous current
    b. If the triggering angle is changed to 120°, calculate the angle at the maximum instantaneous current
    c. Calculate the conduction period for part b

27. A resistive load of 5 Ω is connected to an ac source of 120 V (rms) through an SCR circuit.

a. If the SCR circuit consists of a single SCR, and if the triggering angle is adjusted to 30°, calculate the power consumption of the load

b. If the SCR circuit consists of two, anti-parallel, SCRs, calculate the power consumption of the load assuming that the triggering angle is kept at 30°

28. An inductive load that has a resistive component of 4 Ω is connected to an ac source of 120 V (rms) through a half-wave SCR circuit. When the triggering angle of the SCR is 50°, the conduction period is 160°. Calculate the following:

a. Average voltage across the load

b. Average voltage across the resistive element of the load

c. rms voltage across the load

d. Average current of the load

e. If a freewheeling diode is connected across the load, calculate the load rms voltage. Assume that the current of the diode flows for a complete half cycle

29. The full-wave ac/dc converter is operating under continuous current condition. The source voltage is 120 V (rms) and the load resistance is 2 Ω. For an average load current of 40 A, calculate the triggering angle of the SCRs.

30. A three-phase ac/dc converter is excited by a three-phase source of 480 V (rms and line-to-line). Compute the following:

a. Rms voltage across the load when the triggering angle is 30°

b. Average voltage across the load when the triggering angle is 140°. Keep in mind that the conduction is incomplete when the triggering angle is greater than 90°

6

Induction Generator

Historically, induction machines were mainly used as motors. Since its invention in the nineteenth century by Tesla, they have been the workhorses of the industry. This is because of their ruggedness and low maintenance. As a matter of fact, the three-phase system was invented to create rotating magnetic field that would spin the induction motor. Until the early 1970s, the induction generator was not commonly used because of its lack of excitation circuit. Thus, the generator is hard to use as a standalone system as it relies on external power sources to provide the needed magnetic field. In addition, the machine is hungry for reactive powers that would have to come from an external power source.

The induction machine comes in two main designs: squirrel-cage and wound-rotor. The squirrel-cage design has its rotor windings made of heavy bars that are short circuited and cannot be accessed. The wound-rotor (also known as "slip-ring") consists of three-phase windings that can be accessed from a stationary frame by a system of brushes and slip rings. This way, the rotor can be excited or its windings can be modified by external circuits.

The induction machine is used as a generator for wind turbines; almost all of types 1, 2, and 3 turbines use induction generators. In this chapter, the model for the induction generator is developed for steady-state and dynamic operations.

6.1 Description of Induction Machine

The induction machine consists of a three-phase stator windings as well as three-phase rotor windings. The machines with inaccessible rotor (squirrel-cage) are mainly used in type 1 turbines. The machines with accessible rotor circuit (slip-ring or wound-rotor) are used in types 2 and 3 systems. The stator windings of either machines is connected to a three-phase source to create a rotating magnetic field in the airgap between the stator and rotor that carries the energy from the stator to rotor in case of motor action, or from rotor to stator for generator action.

The schematic of the rotor of the slip-ring generator is shown in Figure 6.1. The three rotor windings of the generator are connected to three slip rings. A brush mechanism consisting of carbon stick attached to a spring keeps the slip ring in contact with the carbon stick while the rotor is spinning. By this system, external circuits on the stationary frame can be in contact with the rotor windings.

Figure 6.2 shows a conceptual diagram of the three-phase stator windings. The figure shows three windings mounted symmetrically inside a hollow metal tube called "stator." The axes of the windings are shifted from each other by 120°. Each winding wraps along the length of the stator on one side and then returns from the other side. If we apply

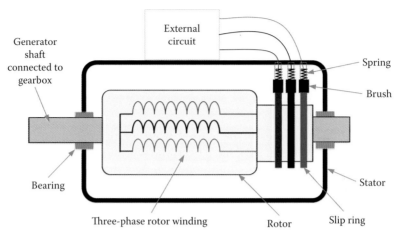

FIGURE 6.1
Main rotor components of a slip-ring induction generator.

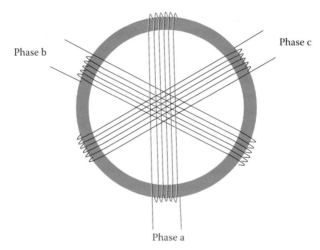

FIGURE 6.2
Three-phase windings mounted on a stator.

a three-phase balanced voltage across the terminals of the windings, the currents of the windings create balanced three-phase magnetic fields inside the tube, as shown in Figure 6.3.

Figure 6.4 shows a more convenient representation of the stator windings. The circles embedded inside the stator represent the windings of the motor: a–a' represent the winding of phase a, b–b' represent the winding of phase b, and c–c' represent the winding of phase c. The crosses and dots inside the circles represent current entering and leaving the windings, respectively.

Because of the mechanical arrangement of the stator windings, the flux of each phase, according to the right-hand rule, travels along its axis, as shown in Figure 6.4. This flux is inside the tube, thus it is called "airgap flux."

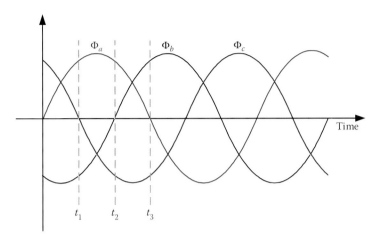

FIGURE 6.3
Airgap flux of the three phases.

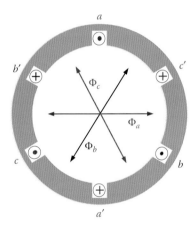

FIGURE 6.4
Loci of airgap fluxes.

If we assume that the current in the coils are balanced, the flux is also balanced. Hence,

$$\Phi_a = \Phi_{max} \sin \omega t$$

$$\Phi_b = \Phi_{max} \sin(\omega t - 120°)$$ (6.1)

$$\Phi_c = \Phi_{max} \sin(\omega t + 120°)$$

Now, let us consider any three consecutive time instances (e.g., t_1, t_2, and t_3 shown in Figure 6.3). At t_1, the angle is 60°, and the magnitude of the flux of each phase is

$$\Phi_a = -\Phi_b = \frac{\sqrt{3}}{2}\Phi_{max}$$ (6.2)

$$\Phi_c = 0$$

The fluxes at t_1 are mapped along their corresponding axes in Figure 6.5a. Note the direction of the current in each winding as it determines the direction of its flux according to the right-hand rule. The total airgap flux is the phasor sum of all fluxes present in the airgap. Hence, at t_1

$$\overline{\Phi}_{total}(t_1) = \overline{\Phi}_a(t_1) + \overline{\Phi}_b(t_1) + \overline{\Phi}_c(t_1)$$

$$= \frac{\sqrt{3}}{2}\Phi_{max}\angle 0° + \frac{\sqrt{3}}{2}\Phi_{max}\angle 60° + 0 \qquad (6.3)$$

$$= \frac{3}{2}\Phi_{max}\angle 30°$$

Similarly, at t_2, the angle is 120°, and the magnitudes of the fluxes of the three phases are

$$\Phi_a = -\Phi_c = \frac{\sqrt{3}}{2}\Phi_{max} \qquad (6.4)$$

$$\Phi_b = 0$$

These fluxes are shown in Figure 6.5b, where the total airgap flux at t_2 is

$$\overline{\Phi}_{total}(t_2) = \overline{\Phi}_a(t_2) + \overline{\Phi}_b(t_2) + \overline{\Phi}_c(t_2)$$

$$= \frac{\sqrt{3}}{2}\Phi_{max}\angle 0° + 0 + \frac{\sqrt{3}}{2}\Phi_{max}\angle -60° \qquad (6.5)$$

$$= \frac{3}{2}\Phi_{max}\angle -30°$$

Finally, at t_3, the angle is 180°, and the magnitudes of the fluxes of the three phases are

$$\Phi_a = 0$$

$$\qquad (6.6)$$

$$\Phi_b = -\Phi_c = \frac{\sqrt{3}}{2}\Phi_{max}$$

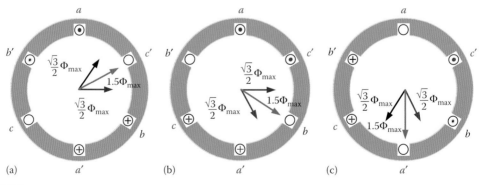

FIGURE 6.5
Rotating airgap flux: (a) t_1, (b) t_2, and (c) t_3.

These fluxes are shown in Figure 6.5c. In this case, the total airgap flux at t_3 is

$$\overline{\Phi}_{total}(t_3) = \overline{\Phi}_a(t_3) + \overline{\Phi}_b(t_3) + \overline{\Phi}_c(t_3)$$

$$= 0 + \frac{\sqrt{3}}{2}\Phi_{max} \angle -120° + \frac{\sqrt{3}}{2}\Phi_{max} \angle -60° \tag{6.7}$$

$$= \frac{3}{2}\Phi_{max} \angle -90°$$

Equations 6.3, 6.5, and 6.7 show that the magnitude of the total flux in the airgap is constant and equal to $(3/2)\Phi_{max}$, the angle of the total airgap flux changes with time, and the total flux in the airgap completes one revolution in every ac cycle. When the grid voltage is balanced and its frequency is constant, the magnetic field is rotating in the airgap with constant magnitude and speed.

The mechanical speed of the flux in the airgap is known as the "synchronous speed," n_s, where

$$n_s = f \tag{6.8}$$

where:
  f is the frequency of the grid (Hz)
  n_s is the synchronous speed in revolution per second

The mechanical speed is often measured in revolution per minute (rpm). Hence, n_s in rpm is

$$n_s = 60f \tag{6.9}$$

The machine in Figure 6.4 is considered to have two poles because every phase has one winding that creates one north pole and one south pole. If each phase is composed of two windings (i.e., a_1–a_1' and a_2–a_2') arranged symmetrically along the inner circumference of the stator, as shown in Figure 6.6, the machine is considered to be four-pole. In this arrangement, the mechanical angle between the phases is 60° instead of the 120° for the two-pole machine. If you repeat the analyses in Equations 6.2 through 6.7, you will find that the flux

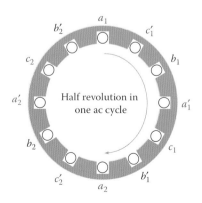

FIGURE 6.6
Four-pole arrangement.

moves 180° (mechanical angle) for each complete ac cycle. Hence, we can write a general expression for the mechanical synchronous speed of the airgap flux as

$$n_s = \frac{60f}{pp} = 120\frac{f}{p}$$ (6.10)

where:
  pp is the number of pole pairs
  p is the number of poles ($p = 2\,pp$)

For the generator operation, the power conversion from mechanical to electrical can be explained using Faraday's law and Lorentz equation. When a conductor moves in a uniform magnetic field, the electromechanical relationships can be represented by the following equations:

$$\text{force} = Bli$$

$$e = Bl\,\Delta v$$ (6.11)

where:
  force is the mechanical force exerted on the conductor
  B is the flux density
  l is the length of the conductor
  i is the current in the conductor
  Δv is the relative speed between the conductor and the flux
  e is the induced voltage across the conductor

If we generalize Equation 6.11 for rotating fields, we can rewrite them in the following forms:

$$T = f(\Phi, i)$$ (6.12)

$$e = f(\Phi, \Delta n)$$ (6.13)

where:
  $f(.)$ is functional relationship
  Φ is the flux
  T is the torque of the conductors
  Δn is the difference in angular speed between the conductor and the flux in the airgap

To understand the mechanism by which the generator operates, keep in mind that the airgap flux is the carrier of energy from the rotor to the stator in the generator operation, and from the stator to the rotor in the motor operation. This airgap flux does exist if the stator is connected to the grid. Assume that the blades of a wind turbine are delivering mechanical power to the rotor of the generator. This mechanical power has a torque T and speed n. The acquired torque produces current in the short-circuited rotor windings according to Equations 6.12. The speed difference Δn in Equations 6.13 is the difference between the rotor speed n and the speed of the airgap flux n_s.

$$\Delta n = n_s - n \tag{6.14}$$

If the rotor speed is higher than the synchronous speed, a negative voltage is induced in the rotor windings according to Equations 6.13. Hence, the induced voltage is a source of energy. The developed power in the rotor windings due to the rotor current and induced voltage is transferred to the stator through the airgap flux. This is the generator operation.

According to Equations 6.13, if the speed of the rotor is less than the synchronous speed, the induced voltage in the rotor is positive and it is a sink of energy. The machine in this case operates as a motor; the power comes to the rotor from the stator through the airgap flux and is delivered to the mechanical load.

As you may conclude from Equations 6.12 and 6.13, if the induction machine is spinning at the synchronous speed, $\Delta n = 0$ and $e = 0$. Thus, no power is developed in the rotor windings. But what if the blades are acquiring power from the wind? In this case, the mechanical power from the blades increases the kinetic energy of the rotating mass causing the speed to increase until the powers are balance; the mechanical power from the blades is equal to the electrical power delivered to the grid plus all losses.

The per unit value of the speed difference is known as slip s.

$$s = \frac{\Delta n}{n_s} = \frac{n_s - n}{n_s} \tag{6.15}$$

If we use the angular speed ω instead of n, the slip is

$$s = \frac{\Delta \omega}{\omega_s} = \frac{\omega_s - \omega}{\omega_s} \tag{6.16}$$

Keep in mind that the unit of n is rpm and that for ω is radian per second (rad/s). Hence,

$$\omega = \frac{2\pi}{60} n \tag{6.17}$$

6.2 Representation of Induction Machine

The induction machine can be represented by six coils: three in the stator and three in the rotor. The axes of all these coils are shown in Figure 6.7. The stator coils are labeled a_1, b_1, and c_1. The rotor coils are a_2, b_2, and c_2. In the next analysis, we use the axis of coil a_1 as the reference frame, which is stationary. The rotor is spinning at speed ω_2, and the angle between the reference and rotor frames is θ_2.

Figure 6.8 shows the circuit representation of the windings of the induction machine. In the following analyses, we shall use the current convention shown in the figures, where all currents are going into the windings. With the six windings, we have 36 inductances representing all possible combinations.

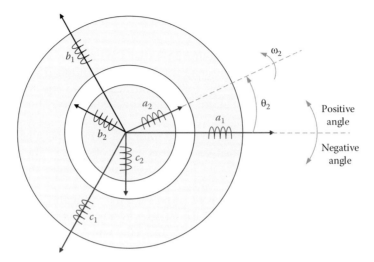

FIGURE 6.7
Windings of the induction machine.

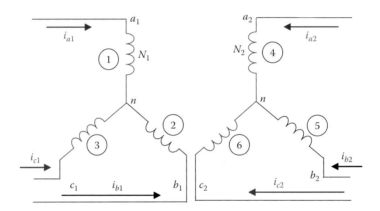

FIGURE 6.8
Circuit representation of the induction machine windings.

$$
\boldsymbol{\ell} =
\begin{bmatrix}
\ell_{11} & \ell_{12} & \ell_{13} & \ell_{14} & \ell_{15} & \ell_{16} \\
\ell_{21} & \ell_{22} & \ell_{23} & \ell_{24} & \ell_{25} & \ell_{26} \\
\ell_{31} & \ell_{32} & \ell_{33} & \ell_{34} & \ell_{35} & \ell_{36} \\
\ell_{41} & \ell_{42} & \ell_{43} & \ell_{44} & \ell_{45} & \ell_{46} \\
\ell_{51} & \ell_{52} & \ell_{53} & \ell_{54} & \ell_{55} & \ell_{56} \\
\ell_{61} & \ell_{62} & \ell_{63} & \ell_{64} & \ell_{65} & \ell_{66}
\end{bmatrix}
\tag{6.18}
$$

where:

$\ell_{xy} = \ell_{yx}$ is the mutual-inductance between windings x and y

ℓ_{xx} is the self-inductance of winding x

Because the thickness of the airgap is constant, the inductances can be grouped into two types: constant and position-dependent inductances. The constant inductances are as follows:

- Self-inductance of any stator winding (L_{s1})
- Self-inductance of any rotor winding (L_{s2})
- Mutual inductance between any two windings in the stator (L_{m1})
- Mutual inductance between any two windings in the rotor (L_{m2})

The position-dependent inductance is the mutual inductance between any stator and rotor windings. This mutual inductance depends on the alignment of their axes. The maximum mutual inductance occurs when the two axes are aligned. When the windings are 90° apart, the mutual inductance is at its minimum value.

Accordingly, the system inductance matrix in Equation 6.18 can be modified as

$$\ell = \begin{bmatrix} L_{s1} & L_{m1} & L_{m1} & \ell_{14} & \ell_{15} & \ell_{16} \\ L_{m1} & L_{s1} & L_{m1} & \ell_{24} & \ell_{25} & \ell_{26} \\ L_{m1} & L_{m1} & L_{s1} & \ell_{34} & \ell_{35} & \ell_{36} \\ \ell_{14} & \ell_{24} & \ell_{34} & L_{s2} & L_{m2} & L_{m2} \\ \ell_{15} & \ell_{25} & \ell_{35} & L_{m2} & L_{s2} & L_{m2} \\ \ell_{16} & \ell_{26} & \ell_{36} & L_{m2} & L_{m2} & L_{s2} \end{bmatrix} \tag{6.19}$$

where:

L_{s1} is the self-inductance of any stator winding

L_{s2} is the self-inductance of any rotor winding

L_{m1} is the mutual inductance between any two stator windings

L_{m2} is the mutual inductance between any two rotor windings

Now let us consider the mutual inductance between the stator and rotor windings. Consider ℓ_{14}, which is the mutual inductance between windings a_1 and a_2.

$$\ell_{14} = L_m \cos(\theta_2) \tag{6.20}$$

where:

L_m is the maximum mutual inductance between a_1 and a_2 when the two windings are aligned

Similarly, ℓ_{24}, which is the mutual inductance between windings b_1 and a_2 is

$$\ell_{24} = L_m \cos(\theta_2 - 120) \tag{6.21}$$

Repeating the process for all other mutual inductances yields.

$$\ell = \begin{bmatrix} L_{s1} & L_{m1} & L_{m1} \\ L_{m1} & L_{s1} & L_{m1} \\ L_{m1} & L_{m1} & L_{s1} \\ L_m\cos(\theta_2) & L_m\cos(\theta_2-120) & L_m\cos(\theta_2+120) \\ L_m\cos(\theta_2+120) & L_m\cos(\theta_2) & L_m\cos(\theta_2-120) \\ L_m\cos(\theta_2-120) & L_m\cos(\theta_2+120) & L_m\cos(\theta_2) \end{bmatrix}$$

$$\begin{matrix} L_m\cos(\theta_2) & L_m\cos(\theta_2+120) & L_m\cos(\theta_2-120) \\ L_m\cos(\theta_2-120) & L_m\cos(\theta_2) & L_m\cos(\theta_2+120) \\ L_m\cos(\theta_2+120) & L_m\cos(\theta_2-120) & L_m\cos(\theta_2) \\ L_{s2} & L_{m2} & L_{m2} \\ L_{m2} & L_{s2} & L_{m2} \\ L_{m2} & L_{m2} & L_{s2} \end{matrix} \quad (6.22)$$

6.2.1 Flux Linkage

To develop a model for the induction machine, we need to find the relationships between the voltages and currents. The first step is to find the flux linkage λ of the induction machine.

$$\lambda = \ell i \qquad (6.23)$$

$$\begin{bmatrix} \lambda_{a1} \\ \lambda_{b1} \\ \lambda_{c1} \\ \lambda_{a2} \\ \lambda_{b2} \\ \lambda_{c2} \end{bmatrix} = \begin{bmatrix} L_{s1} & L_{m1} & L_{m1} \\ L_{m1} & L_{s1} & L_{m1} \\ L_{m1} & L_{m1} & L_{s1} \\ L_m\cos(\theta_2) & L_m\cos(\theta_2-120) & L_m\cos(\theta_2+120) \\ L_m\cos(\theta_2+120) & L_m\cos(\theta_2) & L_m\cos(\theta_2-120) \\ L_m\cos(\theta_2-120) & L_m\cos(\theta_2+120) & L_m\cos(\theta_2) \end{bmatrix}$$

$$\begin{matrix} L_m\cos(\theta_2) & L_m\cos(\theta_2+120) & L_m\cos(\theta_2-120) \\ L_m\cos(\theta_2-120) & L_m\cos(\theta_2) & L_m\cos(\theta_2+120) \\ L_m\cos(\theta_2+120) & L_m\cos(\theta_2-120) & L_m\cos(\theta_2) \\ L_{s2} & L_{m2} & L_{m2} \\ L_{m2} & L_{s2} & L_{m2} \\ L_{m2} & L_{m2} & L_{s2} \end{matrix} \begin{bmatrix} i_{a1} \\ i_{b1} \\ i_{c1} \\ i_{a2} \\ i_{b2} \\ i_{c2} \end{bmatrix} \quad (6.24)$$

The voltage is the derivative of the flux linkage

$$v = \frac{d}{dt}\lambda = \frac{d}{dt}\ell i = \ell\frac{d}{dt}i + \left(\frac{d}{dt}\ell\right)i \qquad (6.25)$$

where:

v is a vector of windings' voltages

i is a vector of windings' currents

Because Equation 6.22 is full matrix and time varying as it is dependent on the rotor position, the analysis of Equation 6.25 is difficult. To simplify the calculations, we can assume a balanced system and use rotating reference frame instead of the stationary reference frame of the stator. These two processes are detailed in the following subsections.

6.2.2 Balanced System

It is reasonable to assume that the induction machine is connected to a balanced three-phase system. In this case, the voltage and currents of the stator voltages and currents are as follows:

$$\begin{bmatrix} v_{a1} \\ v_{b1} \\ v_{c1} \end{bmatrix} = V_{max1} \begin{bmatrix} \sin \omega t \\ \sin(\omega t - 120) \\ \sin(\omega t + 120) \end{bmatrix}$$

$$\begin{bmatrix} i_{a1} \\ i_{b1} \\ i_{c1} \end{bmatrix} = I_{max1} \begin{bmatrix} \sin(\omega t + \phi_1) \\ \sin(\omega t + \phi_1 - 120) \\ \sin(\omega t + \phi_1 + 120) \end{bmatrix}$$

(6.26)

where:

ω is the electrical angular speed ($2\pi f$), where f is the frequency of the supply voltage

ϕ_1 is the power factor angle of the stator current

Because the system is balanced, the phasor sum of all three currents is zero.

$$i_{a1} + i_{b1} + i_{c1} = 0 \tag{6.27}$$

Using the relationship in Equation 6.27, we can simplify Equation 6.24. For instance, consider the flux linkage of phase a_1.

$$\lambda_{a1} = L_{s1}i_{a1} + L_{m1}i_{b1} + L_{m1}i_{c1} + \ell_{14}i_{a2} + \ell_{15}i_{b2} + \ell_{16}i_{c2} \tag{6.28}$$

$$\lambda_{a1} = L_{s1}i_{a1} + L_{m1}\left(i_{b1} + i_{c1}\right) + \ell_{14}i_{a2} + \ell_{15}i_{b2} + \ell_{16}i_{c2} \tag{6.29}$$

$$\lambda_{a1} = L_{s1}i_{a1} - L_{m1}i_{a1} + \ell_{14}i_{a2} + \ell_{15}i_{b2} + \ell_{16}i_{c2} \tag{6.30}$$

$$\lambda_{a1} = \left(L_{s1} - L_{m1}\right)i_{a1} + \ell_{14}i_{a2} + \ell_{15}i_{b2} + \ell_{16}i_{c2} \tag{6.31}$$

$$\lambda_{a1} = L_{11}i_{a1} + \ell_{14}i_{a2} + \ell_{15}i_{b2} + \ell_{16}i_{c2} \tag{6.32}$$

Repeating the process for all other phases yield

$$
\begin{bmatrix} \lambda_{a1} \\ \lambda_{b1} \\ \lambda_{c1} \\ \lambda_{a2} \\ \lambda_{b2} \\ \lambda_{c2} \end{bmatrix} =
\begin{bmatrix}
L_{11} & 0 & 0 \\
0 & L_{11} & 0 \\
0 & 0 & L_{11} \\
L_m\cos(\theta_2) & L_m\cos(\theta_2-120) & L_m\cos(\theta_2+120) \\
L_m\cos(\theta_2+120) & L_m\cos(\theta_2) & L_m\cos(\theta_2-120) \\
L_m\cos(\theta_2-120) & L_m\cos(\theta_2+120) & L_m\cos(\theta_2)
\end{bmatrix}
$$

$$
\begin{bmatrix}
L_m\cos(\theta_2) & L_m\cos(\theta_2+120) & L_m\cos(\theta_2-120) \\
L_m\cos(\theta_2-120) & L_m\cos(\theta_2) & L_m\cos(\theta_2+120) \\
L_m\cos(\theta_2+120) & L_m\cos(\theta_2-120) & L_m\cos(\theta_2) \\
L_{22} & 0 & 0 \\
0 & L_{22} & 0 \\
0 & 0 & L_{22}
\end{bmatrix}
\begin{bmatrix} i_{a1} \\ i_{b1} \\ i_{c1} \\ i_{a2} \\ i_{b2} \\ i_{c2} \end{bmatrix}
$$

$$(6.33)$$

where:

$L_{11} = L_{s1} - L_{m1}$

$L_{22} = L_{s2} - L_{m2}$

$\theta_2 = \omega_2 t$

ω_2 is the rotor's angular speed

Note that the inductance matrix in Equation 6.33 is less crowded as compared with the one in Equation 6.24. The matrix, however, is still time varying. This is addressed in the following subsections.

6.2.3 Rotating Reference Frame

Consider the rotating vector A in Figure 6.9. The figure shows the vector at two times t_1 and t_2. If we compute the projection of this vector on a fixed frame, the projection changes with time. However, if we allow the frame to rotate at the same speed as vector A, the projection is always the same, as shown in Figure 6.10.

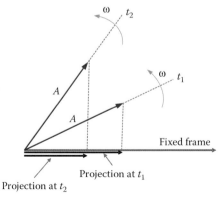

FIGURE 6.9

Projection of a rotating vector on a fixed frame.

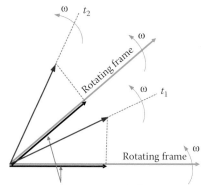

FIGURE 6.10
Projection of a rotating vector on a rotating frame.

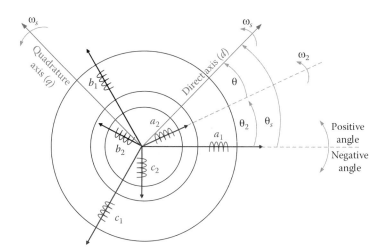

FIGURE 6.11
Induction machine with rotating frames.

Blondel and Robert Park have used this concept to reduce the complexity of the inductance matrix. The main idea is to assume fictitious rotating frames called "direct" and "quadrature," as shown in Figure 6.11. The quadrature axis leads the direct axis by 90°. Moreover, both axes are rotating at the synchronous speed of the magnetic field of the airgap.

The projection of the stator voltage on the direct axis is

$$v_{d1} = K[v_{a1}\cos\theta_s + v_{b1}\cos(\theta_s - 120) + v_{c1}\cos(\theta_s + 120)] \tag{6.34}$$

where:
  K is a scale factor equal to 2/3, which account for converting three-phase quantity into two phases
  $\theta_s = \omega_s t$
  ω_s is the synchronous speed of the magnetic flux

Similarly, the quadrature axis voltage is

$$v_{q1} = -K[v_{a1}\sin\theta_s + v_{b1}\sin(\theta_s - 120) + v_{c1}\sin(\theta_s + 120)] \tag{6.35}$$

In matrix form, we can write the voltage in the direct-quadrature (d-q) frames as a function of the stator voltages

$$\begin{bmatrix} v_{d1} \\ v_{q1} \\ v_{o1} \end{bmatrix} = \frac{2}{3}\begin{bmatrix} \cos\theta_s & \cos(\theta_s - 120) & \cos(\theta_s + 120) \\ -\sin\theta_s & -\sin(\theta_s - 120) & -\sin(\theta_s + 120) \\ 1/2 & 1/2 & 1/2 \end{bmatrix}\begin{bmatrix} v_{a1} \\ v_{b1} \\ v_{c1} \end{bmatrix} \tag{6.36}$$

v_{o1} is introduced to make the matrix square and invertible. Its value is zero for balanced systems.

$$\bar{v}_{o1} = \frac{1}{3}(\bar{v}_{a1} + \bar{v}_{b1} + \bar{v}_{c1}) = 0 \tag{6.37}$$

v_{o1} is known as the "zero sequence voltage."
  Equation 6.36 can be written in the general matrix equation as

$$v_{\text{dqo1}} = B_1 v_{\text{abc1}} \tag{6.38}$$

where:

$$v_{\text{dqo1}} = \begin{bmatrix} v_{d1} \\ v_{q1} \\ v_{o1} \end{bmatrix} \tag{6.39}$$

$$v_{\text{abc1}} = \begin{bmatrix} v_{a1} \\ v_{b1} \\ v_{c1} \end{bmatrix} \tag{6.40}$$

$$B_1 = \frac{2}{3}\begin{bmatrix} \cos\theta_s & \cos(\theta_s - 120) & \cos(\theta_s + 120) \\ -\sin\theta_s & -\sin(\theta_s - 120) & -\sin(\theta_s + 120) \\ 1/2 & 1/2 & 1/2 \end{bmatrix} \tag{6.41}$$

One of the convenient features of matrix B_1 is that its inverse is similar to its transposed

$$B_1^{-1} = \begin{bmatrix} \cos\theta_s & -\sin\theta_s & 1 \\ \cos(\theta_s - 120) & -\sin(\theta_s - 120) & 1 \\ \cos(\theta_s + 120) & -\sin(\theta_s + 120) & 1 \end{bmatrix} \tag{6.42}$$

The same process can be made for the rotor voltages

$$\begin{bmatrix} v_{d2} \\ v_{q2} \\ v_{o2} \end{bmatrix} = \frac{2}{3}\begin{bmatrix} \cos\theta & \cos(\theta - 120) & \cos(\theta + 120) \\ -\sin\theta & -\sin(\theta - 120) & -\sin(\theta + 120) \\ 1/2 & 1/2 & 1/2 \end{bmatrix}\begin{bmatrix} v_{a2} \\ v_{b2} \\ v_{c2} \end{bmatrix} \tag{6.43}$$

The angle θ in Figure 6.11 is the angle between the rotor and reference frames, which is a function of ω_s and ω_2

$$\frac{d\theta}{dt} = \omega_s - \omega_2 = s\omega_s \tag{6.44}$$

Hence,

$$\theta = (\omega_s - \omega_2)t = s\omega_s t = s\theta_s \tag{6.45}$$

Substituting Equation 6.45 into Equation 6.43 yields

$$\begin{bmatrix} v_{d2} \\ v_{q2} \\ v_{o2} \end{bmatrix} = \frac{2}{3} \begin{bmatrix} \cos(s\theta_s) & \cos(s\theta_s - 120) & \cos(s\theta_s + 120) \\ -\sin(s\theta_s) & -\sin(s\theta_s - 120) & -\sin(s\theta_s + 120) \\ 1/2 & 1/2 & 1/2 \end{bmatrix} \begin{bmatrix} v_{a2} \\ v_{b2} \\ v_{c2} \end{bmatrix} \tag{6.46}$$

The above equation can also be written in the general matrix form

$$v_{dqo2} = B_2 v_{abc2} \tag{6.47}$$

Combining Equation 6.38 and 6.47 yields

$$\begin{bmatrix} v_{dqo1} \\ v_{dqo2} \end{bmatrix} = \begin{bmatrix} B_1 & 0 \\ 0 & B_2 \end{bmatrix} \begin{bmatrix} v_{abc1} \\ v_{abc2} \end{bmatrix} \tag{6.48}$$

or

$$v_{dqo} = B v_{abc} \tag{6.49}$$

where:

$$v_{dqo} = \begin{bmatrix} v_{dqo1} \\ v_{dqo2} \end{bmatrix} \tag{6.50}$$

$$v_{abc} = \begin{bmatrix} v_{abc1} \\ v_{abc2} \end{bmatrix} \tag{6.51}$$

$$B = \begin{bmatrix} B_1 & 0 \\ 0 & B_2 \end{bmatrix} \tag{6.52}$$

The current and flux linkage can be treated in the same way. The relationships are given in the following two equations:

$$i_{dqo} = B i_{abc} \tag{6.53}$$

$$\lambda_{dqo} = B \lambda_{abc} \tag{6.54}$$

The objective of all the above analysis in the rotating frame is to simplify the inductance matrix and make it time invariant. This can be done by further processing Equation 6.54

$$\lambda_{dqo} = B\lambda_{abc} = B\boldsymbol{\ell}_{abc}\boldsymbol{i}_{abc} = B\boldsymbol{\ell}_{abc}B^{-1}\boldsymbol{i}_{dqo} = \boldsymbol{\ell}_{dqo}\boldsymbol{i}_{dqo} \qquad (6.55)$$

where:

$$\boldsymbol{\ell}_{dqo} = B\boldsymbol{\ell}_{abc}B^{-1} = \begin{bmatrix} L_{11} & 0 & 0 & L_{12} & 0 & 0 \\ 0 & L_{11} & 0 & 0 & L_{12} & 0 \\ 0 & 0 & L_{11} & 0 & 0 & 0 \\ L_{12} & 0 & 0 & L_{22} & 0 & 0 \\ 0 & L_{12} & 0 & 0 & L_{22} & 0 \\ 0 & 0 & 0 & 0 & 0 & L_{22} \end{bmatrix} \qquad (6.56)$$

Hence,

$$\begin{bmatrix} \lambda_{d1} \\ \lambda_{q1} \\ \lambda_{o1} \\ \lambda_{d2} \\ \lambda_{q2} \\ \lambda_{o2} \end{bmatrix} = \begin{bmatrix} L_{11} & 0 & 0 & L_{12} & 0 & 0 \\ 0 & L_{11} & 0 & 0 & L_{12} & 0 \\ 0 & 0 & L_{11} & 0 & 0 & 0 \\ L_{12} & 0 & 0 & L_{22} & 0 & 0 \\ 0 & L_{12} & 0 & 0 & L_{22} & 0 \\ 0 & 0 & 0 & 0 & 0 & L_{22} \end{bmatrix} \begin{bmatrix} i_{d1} \\ i_{q1} \\ i_{o1} \\ i_{d2} \\ i_{q2} \\ i_{o2} \end{bmatrix} \qquad (6.57)$$

where:

$L_{12} = 3/2\,L_m$

$L_{11} = L_{s1} - L_{m1}$

$L_{22} = L_{s2} - L_{m2}$

Note that the inductance matrix in Equation 6.56 contains only three parameters that are time invariant. This simple inductance matrix allows Robert Park to develop a simple model for the induction machine.

6.3 Park's Equations

Robert Park developed the model of the machine based on the theory of rotating frame. The objective is to find a relationship between voltage and currents of all windings. The voltage is the rate of change of flux linkage, that is,

$$v_{abc} = \frac{d}{dt}\lambda_{abc} = \frac{d}{dt}\left(B^{-1}\lambda_{dqo}\right) = \left(\frac{d}{dt}B^{-1}\right)\lambda_{dqo} + B^{-1}\left(\frac{d}{dt}\lambda_{dqo}\right) \qquad (6.58)$$

Because,

$$v_{abc} = B^{-1}v_{dqo} \qquad (6.59)$$

Hence,

$$v_{dqo} = B\left(\frac{d}{dt}B^{-1}\right)\lambda_{dqo} + \left(\frac{d}{dt}\lambda_{dqo}\right) \qquad (6.60)$$

Substituting $\boldsymbol{B}$ in Equation 6.52 into 6.60 and removing the zero sequence components yields

$$\begin{bmatrix} v_{d1} \\ v_{q1} \\ v_{d2} \\ v_{q2} \end{bmatrix} = \begin{bmatrix} 0 & -\omega & 0 & 0 \\ \omega & 0 & 0 & 0 \\ 0 & 0 & 0 & -s\omega \\ 0 & 0 & s\omega & 0 \end{bmatrix} \begin{bmatrix} \lambda_{d1} \\ \lambda_{q1} \\ \lambda_{d2} \\ \lambda_{q2} \end{bmatrix} + \frac{d}{dt} \begin{bmatrix} \lambda_{d1} \\ \lambda_{q1} \\ \lambda_{d2} \\ \lambda_{q2} \end{bmatrix} \tag{6.61}$$

where:

$$\omega = 2\pi f$$

where:

f is the frequency of the supply voltage

6.3.1 Steady-State Model

The first term in Equation 6.61 is called "speed voltage," and the second term is called "transformer voltage." For balanced system, the transformer voltage is zero as

$$\frac{d}{dt} \begin{bmatrix} \lambda_{d1} \\ \lambda_{q1} \\ \lambda_{o1} \\ \lambda_{d2} \\ \lambda_{q2} \\ \lambda_{o2} \end{bmatrix} = \frac{d}{dt} \boldsymbol{B} \begin{bmatrix} \lambda_{a1} \\ \lambda_{b1} \\ \lambda_{c1} \\ \lambda_{a2} \\ \lambda_{b2} \\ \lambda_{c2} \end{bmatrix} = \frac{d}{dt} \boldsymbol{B} \begin{bmatrix} \lambda_{\max 1} \cos(\omega t) \\ \lambda_{\max 1} \cos(\omega t - 120) \\ \lambda_{\max 1} \cos(\omega t + 120) \\ \vdots \\ \vdots \\ \vdots \end{bmatrix} = \frac{d}{dt} \begin{bmatrix} \lambda_{\max 1} \\ 0 \\ 0 \\ \vdots \\ \vdots \\ \vdots \end{bmatrix} = \begin{bmatrix} 0 \\ 0 \\ 0 \\ \vdots \\ \vdots \\ \vdots \end{bmatrix} \tag{6.62}$$

Hence,

$$\begin{bmatrix} v_{d1} \\ v_{q1} \\ v_{d2} \\ v_{q2} \end{bmatrix} = \begin{bmatrix} 0 & -\omega & 0 & 0 \\ \omega & 0 & 0 & 0 \\ 0 & 0 & 0 & -s\omega \\ 0 & 0 & s\omega & 0 \end{bmatrix} \begin{bmatrix} \lambda_{d1} \\ \lambda_{q1} \\ \lambda_{d2} \\ \lambda_{q2} \end{bmatrix} \tag{6.63}$$

Substituting λ in Equations 6.55 and 6.56 into Equation 6.63 yields

$$\begin{bmatrix} v_{d1} \\ v_{q1} \\ v_{d2} \\ v_{q2} \end{bmatrix} = \begin{bmatrix} 0 & -\omega L_{11} & 0 & -\omega L_{12} \\ \omega L_{11} & 0 & \omega L_{12} & 0 \\ 0 & -s\omega L_{12} & 0 & -s\omega L_{22} \\ s\omega L_{12} & 0 & s\omega L_{22} & 0 \end{bmatrix} \begin{bmatrix} i_{d1} \\ i_{q1} \\ i_{d2} \\ i_{q2} \end{bmatrix} \tag{6.64}$$

$$\boldsymbol{v}_{dq} = \boldsymbol{L}_s \, \boldsymbol{i}_{dq}$$

where:

$$\boldsymbol{L}_s = \begin{bmatrix} 0 & -\omega L_{11} & 0 & -\omega L_{12} \\ \omega L_{11} & 0 & \omega L_{12} & 0 \\ 0 & -s\omega L_{12} & 0 & -s\omega L_{22} \\ s\omega L_{12} & 0 & s\omega L_{22} & 0 \end{bmatrix} \tag{6.65}$$

$$\mathbf{v}_{dq} = \begin{bmatrix} v_{d1} \\ v_{q1} \\ v_{d2} \\ v_{q2} \end{bmatrix}$$

$$\mathbf{i}_{dq} = \begin{bmatrix} i_{d1} \\ i_{q1} \\ i_{d2} \\ i_{q2} \end{bmatrix}$$

The winding resistances can now be added to Equation 6.64

$$\begin{bmatrix} v_{d1} \\ v_{q1} \\ v_{d2} \\ v_{q2} \end{bmatrix} = \begin{bmatrix} r_1 & 0 & 0 & 0 \\ 0 & r_1 & 0 & 0 \\ 0 & 0 & r_2 & 0 \\ 0 & 0 & 0 & r_2 \end{bmatrix} \begin{bmatrix} i_{d1} \\ i_{q1} \\ i_{d2} \\ i_{q2} \end{bmatrix} + \begin{bmatrix} 0 & -\omega L_{11} & 0 & -\omega L_{12} \\ \omega L_{11} & 0 & \omega L_{12} & 0 \\ 0 & -s\omega L_{12} & 0 & -s\omega L_{22} \\ s\omega L_{12} & 0 & s\omega L_{22} & 0 \end{bmatrix} \begin{bmatrix} i_{d1} \\ i_{q1} \\ i_{d2} \\ i_{q2} \end{bmatrix} \quad (6.66)$$

where:
  r_1 is the resistance of the stator windings
  r_2 is the resistance of the rotor windings

Equation 6.66 represents four windings in the d–q frames, as shown in Figure 6.12. We can now use this model to compute any variable such as torque or power easily. The result can be converted back to the original machine frames when needed. This is shown next.

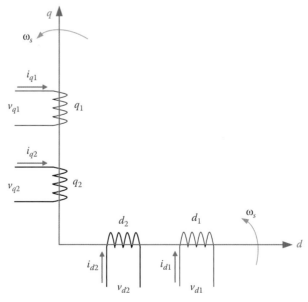

FIGURE 6.12
Equivalent induction machine in a d–q frame.

6.3.1.1 Root Mean Square Values

The relationship of stator voltage in the *abc* frame and that in the *dqo* frame is given in Equation 6.38. This is the same transformation as the currents. Hence,

$$i_{abc1} = B_1^{-1} i_{dqo1} \tag{6.67}$$

$$
\begin{bmatrix} i_{a1} \\ i_{b1} \\ i_{c1} \end{bmatrix} =
\begin{bmatrix}
\cos\omega t & -\sin\omega t & 1 \\
\cos(\omega t - 120) & -\sin(\omega t - 120) & 1 \\
\cos(\omega t + 120) & -\sin(\omega t + 120) & 1
\end{bmatrix}
\begin{bmatrix} i_{d1} \\ i_{q1} \\ i_{o1} \end{bmatrix} \tag{6.68}
$$

Hence,

$$i_{a1} = i_{d1}\cos\omega t - i_{q1}\sin\omega t + i_{o1} \tag{6.69}$$

For a balanced system, the zero sequence current i_{o1} is zero

$$i_{a1} = i_{d1}\cos\omega t - i_{q1}\sin\omega t \tag{6.70}$$

$$i_{a1} = i_{d1}\cos\omega t + i_{q1}\cos(\omega t + 90) \tag{6.71}$$

The two components of the stator current in Equation 6.71 are shown in Figure 6.13. Note that i_{d1} and i_{q1} are the peak values of their own waveforms. Hence, the root mean square (rms) of these currents (I_{d1} and I_{q1}) are

$$
\begin{aligned}
I_{d1} &= \frac{i_{d1}}{\sqrt{2}} \\[2mm]
I_{q1} &= \frac{i_{q1}}{\sqrt{2}}
\end{aligned} \tag{6.72}
$$

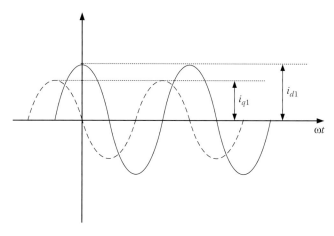

FIGURE 6.13
Waveforms of the direct and quadrature axes currents.

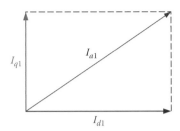

FIGURE 6.14
Stator current as a function of d and q currents.

In phasor form, the current in phase a_1 is

$$\bar{I}_{a1} = I_{d1} + jI_{q1} \tag{6.73}$$

Figure 6.14 is a phasor diagram representing Equation 6.73. The magnitude of the current in phase a_1 is

$$I_{a1} = \sqrt{(I_{d1})^2 + (I_{q1})^2} \tag{6.74}$$

Similarly, we can develop a phasor relationship for the voltage as

$$\bar{V}_{a1} = V_{d1} + jV_{q1} \tag{6.75}$$

6.3.1.2 Real and Reactive Powers

The complex (apparent) power of the machine is the multiplication of all voltages by their corresponding conjugate current. The conjugate is used to make the reactive power of the inductor positive and that for the capacitor negative. This is the norm used by the power community.

The complex power of phase a_1 in the stator.

$$\bar{S}_{a1} = \bar{V}_{a1}\bar{I}_{a1}^* \tag{6.76}$$

For the three phases, the stator power is

$$\bar{S}_1 = 3\bar{S}_{a1} = 3\bar{V}_{a1}\bar{I}_{a1}^* \tag{6.77}$$

Substituting the current and voltage in Equations 6.73 and 6.75 into Equation 6.77

$$\bar{S}_1 = 3(V_{d1} + jV_{q1})(I_{d1} - jI_{q1}) \tag{6.78}$$

The real component of the complex power is the real power

$$P_1 = 3(V_{d1}I_{d1} + V_{q1}I_{q1}) \tag{6.79}$$

And the imaginary component is the reactive power

$$Q_1 = 3(V_{q1}I_{d1} - V_{d1}I_{q1}) \tag{6.80}$$

Similar expressions can be developed for the rotor circuit.

$$P_2 = 3(V_{d2}I_{d2} + V_{q2}I_{q2}) \tag{6.81}$$

$$Q_2 = 3(V_{q2}I_{d2} - V_{d2}I_{q2}) \tag{6.82}$$

6.3.1.3 General Equivalent Circuit

Based on the analyses we have done so far, we can develop an equivalent circuit for the induction machine in the steady-state operation. From Equation 6.66, we get

$$v_{d1} = r_1 i_{d1} - \omega L_{11} i_{q1} - \omega L_{12} i_{q2}$$
$$v_{q1} = r_1 i_{q1} + \omega L_{11} i_{d1} + \omega L_{12} i_{d2} \tag{6.83}$$

In rms quantities,

$$V_{d1} = r_1 I_{d1} - \omega L_{11} I_{q1} - \omega L_{12} I_{q2}$$
$$V_{q1} = r_1 I_{q1} + \omega L_{11} I_{d1} + \omega L_{12} I_{d2} \tag{6.84}$$

The terminal voltage of the stator winding $\overline{V}_{a1}$ is

$$\overline{V}_{a1} = V_{d1} + jV_{q1} = r_1(I_{d1} + jI_{q1}) + \omega L_{11}(jI_{d1} - I_{q1}) + \omega L_{12}(jI_{d2} - jI_{q2})$$
$$\overline{V}_{a1} = V_{d1} + jV_{q1} = r_1(I_{d1} + jI_{q1}) + j\omega L_{11}(I_{d1} + jI_{q1}) + j\omega L_{12}(I_{d2} + jI_{q2}) \tag{6.85}$$

$$\overline{V}_{a1} = V_{d1} + jV_{q1} = r_1\overline{I}_{a1} + j\omega L_{11}\overline{I}_{a1} + j\omega L_{12}\overline{I}_{a2} \tag{6.86}$$

Similarly, the voltage of the rotor winding is

$$\overline{V}_{a2} = V_{d2} + jV_{q2} = r_2\overline{I}_{a2} + js\omega L_{22}\overline{I}_{a2} + js\omega L_{12}\overline{I}_{a1} \tag{6.87}$$

To account for the difference in turns between the stator and rotor, we can refer the rotor variables to the stator windings using the voltage-per-turn and ampere-per-turn rules. The referred rotor voltage to the stator V'_{a2} can be obtained from the voltage-per-turn rule

$$\frac{V'_{a2}}{N_1} = \frac{V_{a2}}{N_2} \tag{6.88}$$

where:
V'_{a2} is the rotor voltage referred to the stator winding
N_1 is the effective number of turns in the stator winding
N_2 is the effective number of turns in the rotor winding

The effective number of turns is the actual number of turns per phase multiplied by a factor to take into account the number of slots per pole per phase, the pitch of the coils, and the change in winding loop length.

The referred rotor current to the stator I'_{a2} can be obtained from the ampere-per-turn rule

$$I'_{a2}N_1 = I_{a2}N_2 \tag{6.89}$$

Rewriting Equation 6.87 using the values in Equations 6.88 and 6.89 yields

$$\frac{N_2}{N_1}\bar{V}'_{a2} = r_2\frac{N_1}{N_2}\bar{I}'_{a2} + js\omega L_{22}\frac{N_1}{N_2}\bar{I}'_{a2} + js\omega L_{12}\bar{I}_{a1} \tag{6.90}$$

$$\bar{V}'_{a2} = r_2\left(\frac{N_1}{N_2}\right)^2\bar{I}'_{a2} + js\omega L_{22}\left(\frac{N_1}{N_2}\right)^2\bar{I}'_{a2} + js\omega L_{12}\bar{I}_{a1} \tag{6.91}$$

Defining

$$r'_2 = r_2\left(\frac{N_1}{N_2}\right)^2$$

$$x_m = \omega L_{12}$$

$$x_1 = \omega(L_{11} - L_{12}) \tag{6.92}$$

$$x'_2 = \omega\left[\left(\frac{N_1}{N_2}\right)^2 L_{22} - L_{12}\right]$$

$$\bar{I}_m = \bar{I}_{a1} + \bar{I}'_{a2}$$

Then Equations 6.86 and 6.90 can be written as

$$\bar{V}_{a1} = r_1\bar{I}_{a1} + jx_1\bar{I}_{a1} + jx_m\bar{I}_m \tag{6.93}$$

$$\frac{\bar{V}'_{a2}}{s} = \frac{r'_2}{s}\bar{I}'_{a2} + jx'_2\bar{I}'_{a2} + jx_m\bar{I}_m \tag{6.94}$$

Equations 6.93 and 6.94 are represented by the equivalent circuit shown in Figure 6.15.

The component r'_2/s can be parsed into two parts: one is the actual rotor resistance and the other is a resistance that is changing with speed.

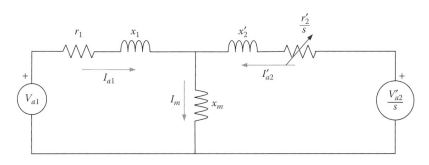

FIGURE 6.15
Equivalent circuit of the induction machine.

$$\frac{r_2'}{s} = r_2' + r_d \tag{6.95}$$

where:

r_d is known as the developed resistance

$$r_d = \frac{r_2'}{s}(1-s) \tag{6.96}$$

The modified equivalent circuit is shown in Figure 6.16.

Note the polarities across r_d in Figure 6.16. Because of the flow of I_{a2}', the positive polarity is when the current enters the resistance. Also, keep in mind that when the machine operates in super-synchronous speed (negative slip), r_d is negative. Consider the conventions of circuit theory shown in Figure 6.17. If the current leaves the positive polarity of a voltage source, the source is delivering energy (discharging). If the current enters the positive polarity of a voltage source, the source is acquiring energy (charged). Also, positive resistance consumes power and negative resistance generates power.

The generic model in Figure 6.16 can be used for induction generator, as shown in Figure 6.18, by considering the following:

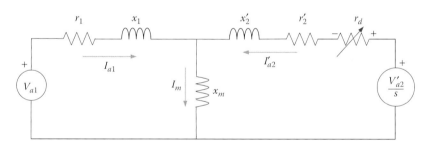

FIGURE 6.16
Generic induction machine model.

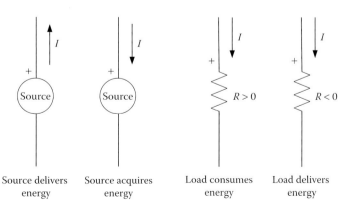

FIGURE 6.17
Circuit theory convention for energy.

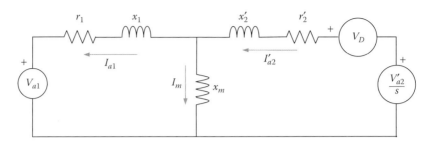

FIGURE 6.18
Alternative model for induction generator.

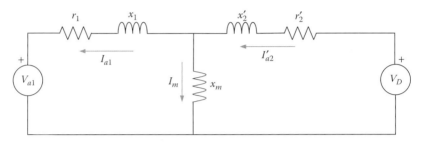

FIGURE 6.19
Induction generator without rotor injection.

- Stator current I_{a1} is reversed as the generator is delivering energy to the grid. This makes V_{a1} in charging mode (receiving energy).

- The voltage across r_d is replaced by a voltage source V_D called "developed voltage." Note that the polarity of V_D indicates a delivery of energy. This voltage source represents the conversion of the mechanical power from the gearbox into electrical power (developed power). This source is delivering energy when r_d is negative (super-synchronous speed). This makes the polarity of V_D, as shown in Figure 6.18, and its magnitude is

$$\overline{V}_D = -r_d\overline{I}'_{a2} = -\frac{r'_2}{s}(1-s)\overline{I}'_{a2} \tag{6.97}$$

For types 1 and 2 systems, the rotor windings are shorted and no external injection is used. In this case, the rotor voltage in the equivalent circuit in Figure 6.18 is eliminated, as shown in Figure 6.19.

The developed power P_d (power coming from the gearbox after subtracting the rotational losses) is

$$P_d = 3V_D I'_{a2} = -3r_d I'^2_{a2} = -3\frac{r'_2}{s}(1-s)I'^2_{a2} \tag{6.98}$$

6.3.1.4 Torque

Mechanical powers can be represented by torque and speed. For the induction generator, we have two mechanical powers that are associated with speed: developed power and airgap power. The developed power is a mechanical power that is converted into electrical. Using mechanical variables (torque and speed), the developed power is

$$P_d = T_d \omega_2 \tag{6.99}$$

where:

T_d is the developed torque; mechanical torque from the gearbox minus friction torque

ω_2 is the speed of the rotor

The power entering the stator is the airgap power P_g, which is carried by the field in the airgap. The airgap flux is spinning at the synchronous speed and its torque is the same as the developed torque. Hence,

$$P_g = T_d \omega_s \tag{6.100}$$

EXAMPLE 6.1

A small type 1 wind turbine has two-pole, 440 V induction generator with following parameters: $r_1 = r_2' = 1.2\,\Omega$; $x_1 = x_2' = 3\,\Omega$; and $x_m = 100\,\Omega$.

The generator is rotating at 1845 rpm. Compute the following:

1. Rotor current referred to stator
2. Stator current
3. Developed power
4. Developed torque
5. Windings' losses
6. Apparent power at the generator terminals
7. Real power at the generator terminals
8. Reactive power consumed by the machine
9. Efficiency of the generator

Solution:

The model in Figure 6.19 can be used to solve this problem. However, since x_m is quite large, we can approximate the equivalent circuit, as shown in Figure 6.20.

1. The slip of the machine is

$$s = \frac{n_s - n}{n_s} = \frac{1800 - 1845}{1800} = -0.025$$

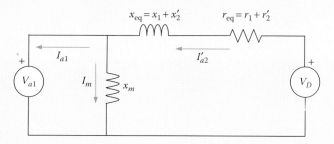

FIGURE 6.20
Approximate equivalent circuit for high magnetizing branch impedance.

The rotor current is

$$\bar{I}'_{a2} = \frac{\bar{V}_D - \bar{V}_{a1}}{r_{eq} + jx_{eq}}$$

Substituting the value of $\bar{V}_D$ from Equation 6.97 yields

$$\bar{I}'_{a2} = \frac{-\bar{V}_{a1}}{(r_{eq} + r_d) + jx_{eq}}$$

where:

$$r_d = \frac{r'_2}{s}(1-s) = \frac{1.2}{-0.025}(1.025) = -49.2$$

Hence,

$$\bar{I}'_{a2} = \frac{-\left(440/\sqrt{3}\right)\angle 0°}{(2.4 - 49.2) + j6} = 5.34 + j0.685 = 5.384\ \angle 7.31°\ A$$

2. To compute the stator current, we need to compute I_m:

$$\bar{I}_m \approx \frac{\bar{V}_{a1}}{jx_m} = \frac{\left(440/\sqrt{3}\right)\angle 0°}{j100} = 2.54\ \angle -90°\ A$$

The stator current is

$$\bar{I}_{a1} = \bar{I}'_{a2} - \bar{I}_m = 5.384\ \angle 7.31° - 2.54\ \angle -90° = 6.24\ \angle 31.13°\ A$$

3. The developed power is

$$P_d = 3V_D I'_{a2} = -3r_d I'^2_{a2} = 3 \times 49.2 \times 5.384^2 = 4.279\ kW$$

4. The developed torque is

$$T_d = \frac{P_d}{\omega_2} = \frac{4.279}{2\pi(1845/60)} = 22.14\ Nm$$

5. The losses in the stator and rotor windings, known as copper losses P_{cu} are

$$P_{cu1} = 3r_1 I^2_{a1} = 3 \times 1.2 \times 6.24^2 = 140.2\ W$$

$$P_{cu2} = 3r_2 I'^2_{a2} = 3 \times 1.2 \times 5.384^2 = 104.35\ W$$

6. The apparent power at the generator terminals is

$$\bar{S} = 3\bar{V}_{a1}\bar{I}^*_{a1} = 3 \times \frac{440}{\sqrt{3}}\ \angle 0° \times 6.24\ \angle -31.13° = 4.756\ \angle -31.13°\ kVA$$

7. The real power delivered to grid is

$$P_s = S \times \cos(31.13°) = 4.756 \times 0.856 = 4.07\ kW$$

8. The reactive power is

$$Q_s = S \times \sin(-31.13°) = -4.756 \times 0.517 = -2.46 \, \text{kVAr}$$

The negative sign in reactive power means that the generator is consuming reactive power from the grid.

9. If we ignore the rotational losses, the efficiency of the generator is

$$\eta = \frac{P_{\text{out}}}{P_{\text{in}}} = \frac{P_s}{P_d} = \frac{4.07}{4.279} = 95.1\%$$

6.3.2 Dynamic Model of Induction Generator

During disturbances or changes in wind speed, the dynamic response of the generator is needed to evaluate the stability of the system as well as to identify unacceptable stress in currents, voltages, torques, and so on. The dynamic models include the electrical and mechanical modes. The electrical modes give information on electrical variables such as voltages and currents. The mechanical modes give information on torques and speeds.

6.3.2.1 Dynamics of Electrical Mode

To develop a dynamic model, we can start with Equation 6.61

$$
\begin{bmatrix} v_{d1} \\ v_{q1} \\ v_{d2} \\ v_{q2} \end{bmatrix} = \begin{bmatrix} 0 & -\omega & 0 & 0 \\ \omega & 0 & 0 & 0 \\ 0 & 0 & 0 & -s\omega \\ 0 & 0 & s\omega & 0 \end{bmatrix} \begin{bmatrix} \lambda_{d1} \\ \lambda_{q1} \\ \lambda_{d2} \\ \lambda_{q2} \end{bmatrix} + \frac{d}{dt} \begin{bmatrix} \lambda_{d1} \\ \lambda_{q1} \\ \lambda_{d2} \\ \lambda_{q2} \end{bmatrix}
\tag{6.101}
$$

If we add the resistances of the windings, we get

$$
\begin{bmatrix} v_{d1} \\ v_{q1} \\ v_{d2} \\ v_{q2} \end{bmatrix} = \begin{bmatrix} r_1 & 0 & 0 & 0 \\ 0 & r_1 & 0 & 0 \\ 0 & 0 & r_2 & 0 \\ 0 & 0 & 0 & r_2 \end{bmatrix} \begin{bmatrix} i_{d1} \\ i_{q1} \\ i_{d2} \\ i_{q2} \end{bmatrix} + \begin{bmatrix} 0 & -\omega & 0 & 0 \\ \omega & 0 & 0 & 0 \\ 0 & 0 & 0 & -s\omega \\ 0 & 0 & s\omega & 0 \end{bmatrix} \begin{bmatrix} \lambda_{d1} \\ \lambda_{q1} \\ \lambda_{d2} \\ \lambda_{q2} \end{bmatrix} + \frac{d}{dt} \begin{bmatrix} \lambda_{d1} \\ \lambda_{q1} \\ \lambda_{d2} \\ \lambda_{q2} \end{bmatrix}
\tag{6.102}
$$

Substituting the flux linkage in Equation 6.57 into the above equation, we get

$$
\begin{bmatrix} v_{d1} \\ v_{q1} \\ v_{d2} \\ v_{q2} \end{bmatrix} = \begin{bmatrix} r_1 & 0 & 0 & 0 \\ 0 & r_1 & 0 & 0 \\ 0 & 0 & r_2 & 0 \\ 0 & 0 & 0 & r_2 \end{bmatrix} \begin{bmatrix} i_{d1} \\ i_{q1} \\ i_{d2} \\ i_{q2} \end{bmatrix} + \begin{bmatrix} 0 & -L_{11}\omega & 0 & -L_{12}\omega \\ L_{11}\omega & 0 & L_{12}\omega & 0 \\ 0 & -L_{12}s\omega & 0 & -L_{22}s\omega \\ L_{12}s\omega & 0 & L_{22}s\omega & 0 \end{bmatrix} \begin{bmatrix} i_{d1} \\ i_{q1} \\ i_{d2} \\ i_{q2} \end{bmatrix}
$$

$$
+ \begin{bmatrix} L_{11} & 0 & L_{12} & 0 \\ 0 & L_{11} & 0 & L_{12} \\ L_{12} & 0 & L_{22} & 0 \\ 0 & L_{12} & 0 & L_{22} \end{bmatrix} \frac{d}{dt} \begin{bmatrix} i_{d1} \\ i_{q1} \\ i_{d2} \\ i_{q2} \end{bmatrix}
\tag{6.103}
$$

Rearranging the above equation, we get

$$
\frac{d}{dt}\begin{bmatrix} i_{d1} \\ i_{q1} \\ i_{d2} \\ i_{q2} \end{bmatrix} = \begin{bmatrix} L_{11} & 0 & L_{12} & 0 \\ 0 & L_{11} & 0 & L_{12} \\ L_{12} & 0 & L_{22} & 0 \\ 0 & L_{12} & 0 & L_{22} \end{bmatrix}^{-1} \begin{bmatrix} -r_1 & L_{11}\omega & 0 & L_{12}\omega \\ -L_{11}\omega & -r_1 & -L_{12}\omega & 0 \\ 0 & L_{12}s\omega & -r_2 & L_{22}s\omega \\ -L_{12}s\omega & 0 & -L_{22}s\omega & -r_2 \end{bmatrix}\begin{bmatrix} i_{d1} \\ i_{q1} \\ i_{d2} \\ i_{q2} \end{bmatrix}
$$

$$
+ \begin{bmatrix} L_{11} & 0 & L_{12} & 0 \\ 0 & L_{11} & 0 & L_{12} \\ L_{12} & 0 & L_{22} & 0 \\ 0 & L_{12} & 0 & L_{22} \end{bmatrix}^{-1} \begin{bmatrix} v_{d1} \\ v_{q1} \\ v_{d2} \\ v_{q2} \end{bmatrix} \tag{6.104}
$$

The inverse of the inductance matrix is

$$
\begin{bmatrix} L_{11} & 0 & L_{12} & 0 \\ 0 & L_{11} & 0 & L_{12} \\ L_{12} & 0 & L_{22} & 0 \\ 0 & L_{12} & 0 & L_{22} \end{bmatrix}^{-1} = \frac{1}{L}\begin{bmatrix} -L_{22} & 0 & L_{12} & 0 \\ 0 & -L_{22} & 0 & L_{12} \\ L_{12} & 0 & -L_{11} & 0 \\ 0 & L_{12} & 0 & -L_{11} \end{bmatrix} \tag{6.105}
$$

where:

$$
L = L_{12}^2 - L_{11}L_{22}
$$

Hence, Equation 6.104 can be written as

$$
\frac{d}{dt}\begin{bmatrix} i_{d1} \\ i_{q1} \\ i_{d2} \\ i_{q2} \end{bmatrix} = \frac{1}{L}\begin{bmatrix} -L_{22} & 0 & L_{12} & 0 \\ 0 & -L_{22} & 0 & L_{12} \\ L_{12} & 0 & -L_{11} & 0 \\ 0 & L_{12} & 0 & -L_{11} \end{bmatrix} \begin{bmatrix} -r_1 & L_{11}\omega & 0 & L_{12}\omega \\ -L_{11}\omega & -r_1 & -L_{12}\omega & 0 \\ 0 & L_{12}s\omega & -r_2 & L_{22}s\omega \\ -L_{12}s\omega & 0 & -L_{22}s\omega & -r_2 \end{bmatrix}\begin{bmatrix} i_{d1} \\ i_{q1} \\ i_{d2} \\ i_{q2} \end{bmatrix}
$$

$$
+ \frac{1}{L}\begin{bmatrix} -L_{22} & 0 & L_{12} & 0 \\ 0 & -L_{22} & 0 & L_{12} \\ L_{12} & 0 & -L_{11} & 0 \\ 0 & L_{12} & 0 & -L_{11} \end{bmatrix}\begin{bmatrix} v_{d1} \\ v_{q1} \\ v_{d2} \\ v_{q2} \end{bmatrix} \tag{6.106}
$$

$$
\frac{d}{dt}\begin{bmatrix} i_{d1} \\ i_{q1} \\ i_{d2} \\ i_{q2} \end{bmatrix} = \frac{1}{L}\begin{bmatrix} r_1 L_{22} & -L_{11}L_{22}\omega + L_{12}^2 s\omega & -r_2 L_{12} & -L_{12}L_{22}\omega_2 \\ L_{11}L_{22}\omega - L_{12}^2 s\omega & r_1 L_{22} & L_{12}L_{22}\omega_2 & -r_2 L_{12} \\ -r_1 L_{12} & L_{11}L_{12}\omega_2 & r_2 L_{11} & L_{12}^2\omega - L_{11}L_{22}s\omega \\ -L_{11}L_{12}\omega_2 & -r_1 L_{12} & -L_{12}^2\omega + L_{11}L_{22}s\omega & r_2 L_{11} \end{bmatrix}\begin{bmatrix} i_{d1} \\ i_{q1} \\ i_{d2} \\ i_{q2} \end{bmatrix}
$$

$$
+ \frac{1}{L}\begin{bmatrix} -L_{22} & 0 & L_{12} & 0 \\ 0 & -L_{22} & 0 & L_{12} \\ L_{12} & 0 & -L_{11} & 0 \\ 0 & L_{12} & 0 & -L_{11} \end{bmatrix}\begin{bmatrix} v_{d1} \\ v_{q1} \\ v_{d2} \\ v_{q2} \end{bmatrix} \tag{6.107}
$$

Equation 6.107 can be simplified to

$$\frac{d}{dt}\begin{bmatrix} i_{d1} \\ i_{q1} \\ i_{d2} \\ i_{q2} \end{bmatrix} = \mathbf{A}_\omega \begin{bmatrix} i_{d1} \\ i_{q1} \\ i_{d2} \\ i_{q2} \end{bmatrix} + \mathbf{B} \begin{bmatrix} v_{d1} \\ v_{q1} \\ v_{d2} \\ v_{q2} \end{bmatrix} \tag{6.108}$$

where:

$$\mathbf{A}_\omega = \frac{1}{L} \begin{bmatrix} r_1 L_{22} & -L_{11}L_{22}\omega + L_{12}^2 s\omega & -r_2 L_{12} & -L_{12}L_{22}\omega_2 \\ L_{11}L_{22}\omega - L_{12}^2 s\omega & r_1 L_{22} & L_{12}L_{22}\omega_2 & -r_2 L_{12} \\ -r_1 L_{12} & L_{11}L_{12}\omega_2 & r_2 L_{11} & L_{12}^2\omega - L_{11}L_{22}s\omega \\ -L_{11}L_{12}\omega_2 & -r_1 L_{12} & -L_{12}^2\omega + L_{11}L_{22}s\omega & r_2 L_{11} \end{bmatrix} \tag{6.109}$$

$$\mathbf{B} = \frac{1}{L} \begin{bmatrix} -L_{22} & 0 & L_{12} & 0 \\ 0 & -L_{22} & 0 & L_{12} \\ L_{12} & 0 & -L_{11} & 0 \\ 0 & L_{12} & 0 & -L_{11} \end{bmatrix} \tag{6.110}$$

The stability of the electrical loop of the generator can be assessed by examining the eigenvalues of $\mathbf{A}_\omega$. A positive real component indicates unstable operation. Note that $\mathbf{A}_\omega$ is a function of the rotor speed of the generator. Hence, the stability can change with the change in the speed of the machine.

6.3.2.2 Rotor Dynamics

Figure 6.21 shows three turbine blades, the net lift force of each blade is computed at the centroid of the blade. These forces cause the turbine to spin in the counterclockwise direction. The net torque of each blade can be computed as

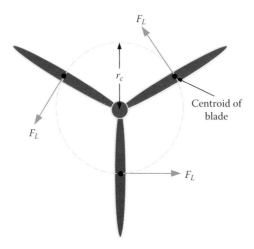

FIGURE 6.21
Lift forces on the turbine blades.

$$T = r_c F_L \tag{6.111}$$

where:

T is the net torque of each blade
F_L is the net lift force of each blade
r_c is the distance from the center of the hub to the centroid of the blade

The total mechanical torque of all blades is

$$T_{\text{blade}} = nT \tag{6.112}$$

where:

T_{blade} is the mechanical torque from all blades
n is the number of blades

The blades' torque is modified by the gearbox according to its speed ratio. For the system in Figure 6.22, the low-speed shaft of the gearbox is spinning at the speed of the blades (ω_{blade}), and its torque is the blade torque (T_{blade}). The high-speed shaft is rotating at the generator speed (ω_2) and the mechanical torque of the blades seen by the generator is T_m. If we assume that the gearbox is lossless, the input and output powers of the gearbox are equal.

$$T_m \omega_2 = T_{\text{blade}} \omega_{\text{blade}} \tag{6.113}$$

Hence, the mechanical torque of the blades seen by the generator is

$$T_m = T_{\text{blade}} \frac{\omega_{\text{blade}}}{\omega_2} \tag{6.114}$$

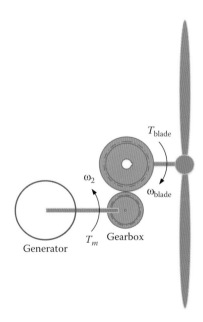

FIGURE 6.22
Speeds and torques due to gearbox.

During the steady-state operation, the speed of the turbine is fairly constant. The mechanical torque from the gearbox (T_m) during the steady state is counteracted by the electric developed torque of the generator (T_d). The developed torque is equal and in opposite direction to T_m. During disturbance, these torques may not be equal, which would cause the speed of the turbine to changes. This can be seen from the torque equation of rotating mass

$$T_m - T_d = 2H \frac{d\omega_2}{dt} \tag{6.115}$$

where:
 H is the inertia constant of the rotating mass (blades, generator, gearbox, etc.)
 $d\omega_2/dt$ is the acceleration of the generator's speed

During steady-state operation, both torques are equal. Thus, the acceleration is zero and the machine operates at constant speed. If the electrical torque (developed torque) is less than the mechanical torque, the generator speeds up, and vice versa.

The developed torque can be computed from the airgap power equation. If we ignore the stator losses, the instantaneous airgap power is

$$p_g = \begin{bmatrix} v_{a1} & v_{b1} & v_{c1} \end{bmatrix} \begin{bmatrix} i_{a1} \\ i_{b1} \\ i_{c1} \end{bmatrix} \tag{6.116}$$

where:

$$i_{abc1} = B_1^{-1} i_{dqo1}$$
$$v_{abc1} = B_1^{-1} v_{dqo1} \tag{6.117}$$

The airgap power can then be written as

$$p_g = (B_1^{-1} v_{dqo1})^T B_1^{-1} i_{dqo1} \tag{6.118}$$

which leads to

$$p_g = \frac{3}{2} \begin{bmatrix} v_{d1} & v_{q1} \end{bmatrix} \begin{bmatrix} i_{d1} \\ i_{q1} \end{bmatrix} \tag{6.119}$$

The direct and quadrature axes voltages in Equation 6.63 assuming a balanced three-phase voltages is

$$\begin{bmatrix} v_{d1} \\ v_{q1} \end{bmatrix} = \begin{bmatrix} 0 & -\omega \\ \omega & 0 \end{bmatrix} \begin{bmatrix} \lambda_{d1} \\ \lambda_{q1} \end{bmatrix} \tag{6.120}$$

Substituting Equation 6.120 into Equation 6.119 yields

$$p_g = \frac{3}{2}(\omega\lambda_{d1}i_{q1} - \omega\lambda_{q1}i_{d1}) \tag{6.121}$$

where:
  $\omega = 2\pi f$, with f as the grid frequency

Keep in mind that the mechanical synchronous speed of the machine is

$$n_s = \frac{120 f}{p} \tag{6.122}$$

where:
  p is the number of poles of the machine
  n_s is the mechanical synchronous speed (rpm)

The angular mechanical speed in rad/sec is

$$\omega_s = 2\pi\frac{n_s}{60} = \frac{2\pi f}{pp} = \frac{\omega}{pp} \tag{6.123}$$

where:
  pp is the number of pole pairs

Substituting the value of ω in Equation 6.123 into Equation 6.121 leads to the airgap power equation in terms of the mechanical speed of the airgap flux.

$$p_g = \frac{3}{2}\frac{pp}{(\omega_s\lambda_{d1}i_{q1} - \omega_s\lambda_{q1}i_{d1})} \tag{6.124}$$

The airgap power is the developed torque multiplied by the speed of the airgap flux ω_s. Hence, the developed torque is

$$T_d = \frac{p_g}{\omega_s} = \frac{3}{2}\frac{pp}{2}(\lambda_{d1}i_{q1} - \lambda_{q1}i_{d1}) \tag{6.125}$$

The values of λ_{d1} and λ_{q1} can be represented by currents as given in Equation 6.57. In this case, we can write the developed torque as a function of currents

$$T_d = \frac{3}{2}\frac{pp}{2}\left[(L_{11}i_{d1} + L_{12}i_{d2})i_{q1} - (L_{11}i_{q1} + L_{12}i_{q2})i_{d1}\right]$$

$$T_d = \frac{3}{2}\frac{pp}{2}L_{12}(i_{d2}i_{q1} - i_{q2}i_{d1}) \tag{6.126}$$

A block diagram representing the rotor dynamics based on the above equations is shown in Figure 6.23. Any change in the mechanical torque will lead to a change in the rotor speed that would eventually change the developed torque to match the change in the mechanical torque. Keep in mind that B and L_s in the figure are time dependent as given in Equations 6.43, 6.46, and 6.64.

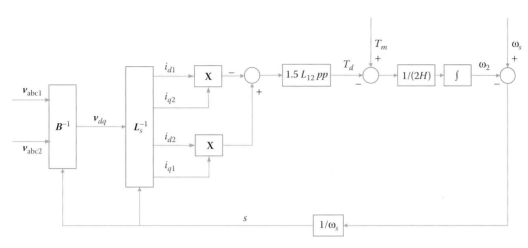

FIGURE 6.23
Dynamic model of an induction generator.

Exercise

1. What are the main components of squirrel-cage induction generator?
2. What are the main components of wound-rotor induction generator?
3. What is the function of the slip rings and brushes?
4. How does magnetic field rotate?
5. What are the main variables determining the speed of rotation of the magnetic field?
6. What are the inductances that change with rotor position?
7. Why is rotating reference frame used to model induction generator?
8. Why induction machine operate with a slip?
9. What is the difference between a developed torque and an airgap torque?
10. An induction generator has the following parameters:

$$L_m = 16 \text{ mH}$$

$$L_{s1} = 30 \text{ mH}$$

$$L_{s2} = 29 \text{ mH}$$

$$L_{m1} = 5 \text{ mH}$$

$$L_{m2} = 4 \text{ mH}$$

Compute all inductive reactance of the equivalent circuit.

11. An induction generator is driven by a wind turbine. The output voltage and current of the generator in rms are $V_{d1} = 600$ V, $V_{q1} = 100$ V, $I_{d1} = 200$ A, and $I_{q1} = 100$ A. Compute the following:

a. Magnitude of the output power of the generator

b. Reactive power at the terminal of the generator

c. Stator current

d. Power factor angle

12. A wind turbine has six poles, 690 V induction generator with the following parameters: $r_1 = r_2' = 0.05\,\Omega$; $x_1 = x_2' = 0.5\,\Omega$; and $x_m = 50\,\Omega$.

The generator is rotating at 1230 rpm. Compute the following:

a. Rotor current referred to stator

b. Stator current

c. Developed power

d. Developed torque

e. Windings' losses

f. Apparent power at the generator terminals

g. Real power at the generator terminals

h. Reactive power consumed by the machine

i. Efficiency of the generator

13. A 690 V induction generator has the following parameters: $x_1 = 2\,\Omega$; $x_2' = 1.5\,\Omega$; $x_m = 100\,\Omega$; $r_1 = 1\,\Omega$; and $r_2' = 1\,\Omega$.

When the slip is −0.1, compute the following:

a. Reactive power of the generator

b. Developed power

c. Power delivered to the grid

d. Power factor at the generator terminals

14. A six-pole, 60 Hz, three-phase induction generator is receiving 500 hp from external mechanical source and is rotating at 1250 rpm. The rotational loss of the generator is 1 kW. Compute the following:

a. Slip of the generator

b. Developed power

15. A wind turbine with eight poles, 60 Hz three-phase, Y-connected induction generator is delivering 500 kW to the grid at 0.7 pf lagging. The terminal voltage of the generator is 690 V. The electrical efficiency of the generator is 90% and the mechanical efficiency is 92%. The parameters of the machine are $r_1 = r_2' = 0.02\,\Omega$ and $x_{eq} = 0.1\,\Omega$.

Compute the following:

a. Mechanical power received from wind

b. Developed power

c. Stator current

16. A wind turbine has 25 m long blades. The far stream wind speed is 15 m/s and the coefficient of performance C_p is 30% at the operational pitch angle. The induction

generator is six-pole, 60 Hz machine and is rotating at 1260 rpm. The mechanical efficiency of the turbine is 85% and the electrical efficiency of the generator is 90%. Compute the following:

a. Power captured by the blade
b. Slip of the generator
c. Power delivered to the stator
d. Output electric power of the wind turbine
e. Rotor copper loss
f. Stator losses
g. Total system efficiency

7

Synchronous Generator

The synchronous generator is the most popular machine used to generate electricity in power plants. Worldwide, over 98% of all electric power is generated by the synchronous generators. The machine has excitation circuit connected to the rotor. For small-sized generators, the excitation is produced by ferrite permanent magnetic material. This is the most economic design for fractional horsepower generators that do not experience repeated surges in stator currents (repeated surges can demagnetize the rotor). In better designs, the rotor is made of rare earth permanent magnetic material such as the samarium–cobalt to produce stronger magnetic fields. One great advantage of generators built with rare earth material is their high power/volume ratios that make them small in size and weight. The strong rare earth permanent magnetic material allows designers to build generators as large as 1.0 MW. Another advantage is that the rare earth permanent magnet cannot be easily demagnetized, so it can be used for applications where heavy currents and surges in currents are expected. Because of these advantages, the rare earth permanent magnetic synchronous generator is used in wind turbine.

For high power production, the field of the generator is electric to produce strong flux intensity; this also allows the operator to control the terminal voltage and reactive power of the generator. This type is used in conventional power plants and some large wind turbines.

7.1 Description of Synchronous Generator

The synchronous generator consists of a stator and a rotor, as shown in Figure 7.1a. The stator of the synchronous machine is similar to the stator of the induction machine; it consists of three-phase windings mounted symmetrically inside the stator. The stator windings are also known as "armature windings." To produce electric field, the winding of the rotor is excited by an external dc source through a slip-ring system. The rotor winding, which is also known as "field winding" or "excitation winding," produces a stationary flux (ϕ_f) with respect to the rotor.

When a wind turbine spins the rotor of the synchronous generator, the magnetic field cuts the stator windings, thus inducing sinusoidal voltages across the stator windings. Because the three stator windings are equally spaced from each other, the induced voltages across the phase windings are shifted by 120° from each other, as shown in Figure 7.1b.

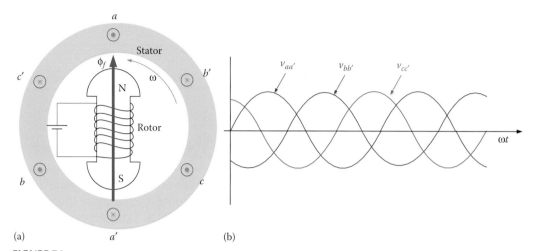

FIGURE 7.1
Three-phase voltage waveforms due to the clockwise rotation of the rotor magnet: (a) main components and (b) induced voltage in stator winding.

The frequency of the induced voltage is dependent on the speed of the rotor. The relationship between the speed of the magnetic field and the frequency of the induced voltage is

$$n = 120\frac{f}{p} \tag{7.1}$$

where:
   n is the speed at which the generator is spinning
   p is the number of poles of the machine
   f is the frequency of the terminal voltage of the generator

There are two types of synchronous generators: cylindrical rotor and salient pole. With cylindrical rotor, the airgap between the stator and rotor is always constant. For the salient rotor, seen in Figure 7.1a, the airgap between any point on the stator surface and the rotor is changing according to the rotor position. The cylindrical rotor type is suitable for high-speed applications such as thermal power plants and the salient pole is more suited for low-speed applications such as hydro power plants and wind.

7.2 Salient Pole Synchronous Generator

The axes of the windings of the synchronous generator in Figure 7.1a are shown in Figure 7.2. Axes a, b, and c are for the stator windings. These axes are stationary. The d axis (direct axis) is aligned with the rotor flux, which is rotating at the speed of the rotor. The q axis (quadrature axis) is 90° leading the direct axis and is also spinning at the speed of the rotor. The angle θ between the axis of phase a and the d axis is changing with time according to the speed of the rotor.

$$\theta = \omega t \tag{7.2}$$

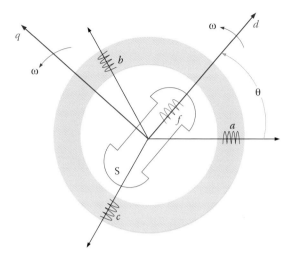

FIGURE 7.2
Axes of a synchronous machine.

The flux linking each winding is the phasor sum of all flux linkage from all windings.

$$
\begin{bmatrix} \lambda_a \\ \lambda_b \\ \lambda_c \\ \lambda_f \end{bmatrix} = \begin{bmatrix} \ell_{aa} & \ell_{ab} & \ell_{ac} & \ell_{af} \\ \ell_{ab} & \ell_{bb} & \ell_{bc} & \ell_{bf} \\ \ell_{ac} & \ell_{bc} & \ell_{cc} & \ell_{cf} \\ \ell_{af} & \ell_{bf} & \ell_{cf} & \ell_{ff} \end{bmatrix} \begin{bmatrix} i_a \\ i_b \\ i_c \\ i_f \end{bmatrix}
\tag{7.3}
$$

where:
  λ_a is the instantaneous flux linkage of phase a
  i_a is the instantaneous current of phase a
  ℓ_{aa} is the self inductance of phase a winding
  ℓ_{ff} is the self inductance of the field winding
  ℓ_{ab} is the mutual inductance between phase a and b windings
  ℓ_{af} is the mutual inductance between phase a and the field windings

Because the airgap distance when measured from the rotor is constant regardless of the rotor position, ℓ_{ff} is not a function of the rotor position, thus it is constant, $\ell_{ff} = L_{ff}$. All other inductances vary with the rotor position. For example, the mutual inductance between the field winding and the stator winding has a maximum value when they are aligned at $\theta = 0$, and zero value when $\theta = 90°$. Hence, the mutual inductances between the rotor and stator windings can be represented by

$$\ell_{af} = L_{af} \cos\theta$$

$$\ell_{bf} = L_{af} \cos(120° - \theta) = L_{af} \cos(\theta - 120°)
\tag{7.4}$$

$$\ell_{cf} = L_{af} \cos(\theta + 120°)$$

where:
  L_{af} is the maximum mutual inductance between the rotor and stator winding at $\theta = 0$

Additionally, the stator self-inductance, such as ℓ_{ab}, is dependent on its flux passing through all possible routes. One of these paths is from a stator winding through the airgap to the rotor and back to the same winding through the airgap. This component is changing with the rotor position. It can be shown that these self inductances of the stator windings are as follows:

$$\ell_{aa} = L_{aa0} + L_g \cos 2\theta$$

$$\ell_{bb} = L_{aa0} + L_g \cos(2\theta + 120°) \tag{7.5}$$

$$\ell_{cc} = L_{aa0} + L_g \cos(2\theta - 120°)$$

$$L_{aa0} = L_{al} + L_{g0}$$

$$L_{g0} = N_a^2 \left(\frac{\mathfrak{R}_q + \mathfrak{R}_d}{2\mathfrak{R}_q \mathfrak{R}_d} \right) \tag{7.6}$$

$$L_g = N_a^2 \left(\frac{\mathfrak{R}_q - \mathfrak{R}_d}{2\mathfrak{R}_q \mathfrak{R}_d} \right)$$

where:
    N_a is a number of turns of the stator winding
    $\mathfrak{R}_d$ is the reluctance seen by the flux crossing the airgap through the direct axis
    $\mathfrak{R}_q$ is the reluctance seen by the flux crossing the airgap through the quadrature axis
    L_{al} is a the leakage inductance due to the flux that does not completely cross the airgap

Finally, the mutual inductance between the stator windings can be derived as

$$\ell_{ab} = -0.5 L_{g0} + L_g \cos(2\theta - 120°)$$

$$\ell_{bc} = -0.5 L_{g0} + L_g \cos 2\theta \tag{7.7}$$

$$\ell_{ca} = -0.5 L_{g0} + L_g \cos(2\theta + 120°)$$

Based on Equations 7.4, 7.5, and 7.7, the inductance matrix in Equation 7.3 can be written as

$$\ell = \begin{bmatrix} L_{aa0} + L_g \cos 2\theta & -0.5 L_{g0} + L_g \cos(2\theta - 120°) & -0.5 L_{g0} + L_g \cos(2\theta + 120°) & L_{af} \cos\theta \\ -0.5 L_{g0} + L_g \cos(2\theta - 120°) & L_{aa0} + L_g \cos(2\theta + 120°) & -0.5 L_{g0} + L_g \cos 2\theta & L_{af} \cos(\theta - 120°) \\ -0.5 L_{g0} + L_g \cos(2\theta + 120°) & -0.5 L_{g0} + L_g \cos 2\theta & L_{aa0} + L_g \cos(2\theta - 120°) & L_{af} \cos(\theta + 120°) \\ L_{af} \cos\theta & L_{af} \cos(\theta - 120°) & L_{ff} \cos(\theta + 120°) & L_{ff} \end{bmatrix} \tag{7.8}$$

7.2.1 Rotating Reference Frame

Because Equation 7.8 is a time-varying full matrix, using the equation is difficult to develop a model for the synchronous generator. The complexity of the matrix is because the stationary axis of phase a of the stator is used as the reference frame. If we instead switch to the rotating axis as our new reference frame, we can obtain an inductance matrix that is much less dependent on the rotor position. This process is similar to what is done in Chapter 6 for the induction machine.

Using Figure 7.2, the projection of the stator currents on the direct axis is

$$i_d = K\left[i_a \cos\theta + i_b \cos(\theta - 120) + i_c \cos(\theta + 120)\right] \tag{7.9}$$

where:

K is a scale factor equal to 2/3, which account for converting three-phase quantity into two phases

$\theta = \omega t$, and ω is the speed of the rotor

Similarly, the quadrature axis voltage is

$$i_q = -K\left[i_a \sin\theta + i_b \sin(\theta - 120) + i_c \sin(\theta + 120)\right] \tag{7.10}$$

In matrix form, we can write the currents in the dq frame as a function of the stator currents.

$$\begin{bmatrix} i_d \\ i_q \\ i_o \\ i_f \end{bmatrix} = \frac{2}{3} \begin{bmatrix} \cos\theta & \cos(\theta - 120) & \cos(\theta + 120) & 0 \\ -\sin\theta & -\sin(\theta - 120) & -\sin(\theta + 120) & 0 \\ 1/2 & 1/2 & 1/2 & 0 \\ 0 & 0 & 0 & 3/2 \end{bmatrix} \begin{bmatrix} i_a \\ i_b \\ i_c \\ i_f \end{bmatrix} \tag{7.11}$$

i_o is the zero-sequence current that is equal to zero for balanced system. It is introduced here to make the matrix square and invertible.

$$\bar{i}_o = \frac{1}{3}(\bar{i}_a + \bar{i}_b + \bar{i}_c) = 0 \tag{7.12}$$

Equation 7.11 can be written in the general matrix equation as

$$\boldsymbol{i}_{\text{dqof}} = \boldsymbol{B}\boldsymbol{i}_{\text{abcf}} \tag{7.13}$$

where:

$$\boldsymbol{i}_{\text{dqof}} = \begin{bmatrix} i_d \\ i_q \\ i_o \\ i_f \end{bmatrix} \tag{7.14}$$

$$\boldsymbol{i}_{\text{abcf}} = \begin{bmatrix} i_a \\ i_b \\ i_c \\ i_f \end{bmatrix} \tag{7.15}$$

$$\boldsymbol{B} = \frac{2}{3} \begin{bmatrix} \cos\theta & \cos(\theta - 120) & \cos(\theta + 120) & 0 \\ -\sin\theta & -\sin(\theta - 120) & -\sin(\theta + 120) & 0 \\ 1/2 & 1/2 & 1/2 & 0 \\ 0 & 0 & 0 & 3/2 \end{bmatrix} \tag{7.16}$$

One of the convenient features of matrix B is that its inverse is similar to its transposed

$$
B^{-1} = \begin{bmatrix} \cos\theta & -\sin\theta & 1 & 0 \\ \cos(\theta-120) & -\sin(\theta-120) & 1 & 0 \\ \cos(\theta+120) & -\sin(\theta+120) & 1 & 0 \\ 0 & 0 & 0 & 1 \end{bmatrix} \tag{7.17}
$$

The voltage and flux linkage can be treated in the same way. The relationships are given in the following two equations:

$$
v_{\text{dqof}} = B v_{\text{abcf}} \tag{7.18}
$$

$$
\lambda_{\text{dqof}} = B \lambda_{\text{abcf}} \tag{7.19}
$$

Equation 7.19 can be written as

$$
\lambda_{\text{dqof}} = B\lambda_{\text{abcf}} = B\ell_{\text{abcf}} i_{\text{abcf}} = B\ell_{\text{abcf}} B^{-1} i_{\text{dqof}} = \ell_{\text{dqof}} i_{\text{dqof}} \tag{7.20}
$$

where:

$$
\ell_{\text{dqof}} = B\ell_{\text{abcf}} B^{-1} = \begin{bmatrix} L_d & 0 & 0 & L_{af} \\ 0 & L_q & 0 & 0 \\ 0 & 0 & L_o & 0 \\ 1.5L_{af} & 0 & 0 & L_{ff} \end{bmatrix} \tag{7.21}
$$

where:

$$
L_d = L_{al} + 3/2(L_{g0} + L_g)
$$

$$
L_q = L_{al} + 3/2(L_{g0} - L_g)
$$

$$
L_o = L_{al}
$$

where:
L_d is the direct axis inductance
L_q is the quadrature axis inductance
L_o is the zero sequence inductance

Equation 7.20 can now be written as

$$
\begin{bmatrix} \lambda_d \\ \lambda_q \\ \lambda_o \\ \lambda_f \end{bmatrix} = \begin{bmatrix} L_d & 0 & 0 & L_{af} \\ 0 & L_q & 0 & 0 \\ 0 & 0 & L_o & 0 \\ 1.5L_{af} & 0 & 0 & L_{ff} \end{bmatrix} \begin{bmatrix} i_d \\ i_q \\ i_o \\ i_f \end{bmatrix} \tag{7.22}
$$

The relationship between the flux linkage and current in Equation 7.22 is much simpler than the one in Equation 7.3. The model development in this chapter is based on Equation 7.22.

7.2.2 Parks Equations

Robert Park developed the relationship between the voltages and currents for all windings in the rotating frame. In his analysis, it is assumed that all currents are injected into the windings (i.e., motor operation). This will be modified later for generator operation. The voltage in the *abcf* frame is

$$v_{abcf} = \frac{d}{dt}\lambda_{abcf} = \frac{d}{dt}\left(B^{-1}\lambda_{dqof}\right) = \left(\frac{d}{dt}B^{-1}\right)\lambda_{dqof} + B^{-1}\left(\frac{d}{dt}\lambda_{dqof}\right) \tag{7.23}$$

Because

$$v_{abcf} = B^{-1}v_{dqof} \tag{7.24}$$

Hence,

$$v_{dqof} = B\left(\frac{d}{dt}B^{-1}\right)\lambda_{dqof} + \left(\frac{d}{dt}\lambda_{dqof}\right) \tag{7.25}$$

Substituting B in Equation 7.16 into 7.25 yields

$$\begin{bmatrix} v_d \\ v_q \\ v_o \\ v_f \end{bmatrix} = \begin{bmatrix} 0 & -\omega & 0 & 0 \\ \omega & 0 & 0 & 0 \\ 0 & 0 & 0 & 0 \\ 0 & 0 & 0 & 0 \end{bmatrix} \begin{bmatrix} \lambda_d \\ \lambda_q \\ \lambda_o \\ \lambda_f \end{bmatrix} + \frac{d}{dt} \begin{bmatrix} \lambda_d \\ \lambda_q \\ \lambda_o \\ \lambda_f \end{bmatrix} \tag{7.26}$$

where:
  $\omega = d\theta/dt$ is the angular speed of the rotor

Adding the voltage drop across the resistances of the windings, Equation 7.26 can be modified as

$$\begin{bmatrix} v_d \\ v_q \\ v_o \\ v_f \end{bmatrix} = \begin{bmatrix} r_a & 0 & 0 & 0 \\ 0 & r_a & 0 & 0 \\ 0 & 0 & r_a & 0 \\ 0 & 0 & 0 & r_f \end{bmatrix} \begin{bmatrix} i_d \\ i_q \\ i_o \\ i_f \end{bmatrix} + \begin{bmatrix} 0 & -\omega & 0 & 0 \\ \omega & 0 & 0 & 0 \\ 0 & 0 & 0 & 0 \\ 0 & 0 & 0 & 0 \end{bmatrix} \begin{bmatrix} \lambda_d \\ \lambda_q \\ \lambda_o \\ \lambda_f \end{bmatrix} + \frac{d}{dt} \begin{bmatrix} \lambda_d \\ \lambda_q \\ \lambda_o \\ \lambda_f \end{bmatrix} \tag{7.27}$$

The first term in Equation 7.27 represents the voltage drop across the resistive components of the machine windings. The second term is called "speed voltage," and the third term is called "transformer voltage."

7.2.3 Steady-State Model

For balanced steady-state system, the transformer voltage in Equation 7.27 is zero.

$$\frac{d}{dt}\begin{bmatrix} \lambda_d \\ \lambda_q \\ \lambda_o \\ \lambda_f \end{bmatrix} = \frac{d}{dt}B\begin{bmatrix} \lambda_a \\ \lambda_b \\ \lambda_c \\ \lambda_f \end{bmatrix} = \frac{d}{dt}B\begin{bmatrix} \lambda_{max}\cos(\omega t) \\ \lambda_{max}\cos(\omega t - 120) \\ \lambda_{max}\cos(\omega t + 120) \\ \lambda_f \end{bmatrix} = \frac{d}{dt}\begin{bmatrix} \lambda_{max} \\ 0 \\ 0 \\ \lambda_f \end{bmatrix} = \begin{bmatrix} 0 \\ 0 \\ 0 \\ d\lambda_f/dt \end{bmatrix} \tag{7.28}$$

For constant field, $d\lambda_f/dt = 0$

Also, for balanced system, the zero sequence current and zero sequence flux are equal to zero. Hence, Equation 7.27 can be modified as

$$\begin{bmatrix} v_d \\ v_q \\ v_f \end{bmatrix} = \begin{bmatrix} r_a & 0 & 0 \\ 0 & r_a & 0 \\ 0 & 0 & r_f \end{bmatrix} \begin{bmatrix} i_d \\ i_q \\ i_f \end{bmatrix} + \begin{bmatrix} 0 & -\omega & 0 \\ \omega & 0 & 0 \\ 0 & 0 & 0 \end{bmatrix} \begin{bmatrix} \lambda_d \\ \lambda_q \\ \lambda_f \end{bmatrix} \tag{7.29}$$

Substituting the value of λ_{dqf} in Equation 7.22 into Equation 7.29 yields

$$\begin{bmatrix} v_d \\ v_q \\ v_f \end{bmatrix} = \begin{bmatrix} r_a & 0 & 0 \\ 0 & r_a & 0 \\ 0 & 0 & r_f \end{bmatrix} \begin{bmatrix} i_d \\ i_q \\ i_f \end{bmatrix} + \begin{bmatrix} 0 & -\omega L_q & 0 \\ \omega L_d & 0 & \omega L_{af} \\ 0 & 0 & 0 \end{bmatrix} \begin{bmatrix} i_d \\ i_q \\ i_f \end{bmatrix} \tag{7.30}$$

7.2.3.1 Root Mean Square Values

The relationship of the currents in the *abcf* frame and that in the *dqof* frame is given in Equation 7.13 as follows:

$$i_{abcf} = B^{-1} i_{dqof} \tag{7.31}$$

$$\begin{bmatrix} i_a \\ i_b \\ i_c \\ i_f \end{bmatrix} = \begin{bmatrix} \cos\theta & -\sin\theta & 1 & 0 \\ \cos(\theta-120) & -\sin(\theta-120) & 1 & 0 \\ \cos(\theta+120) & -\sin(\theta+120) & 1 & 0 \\ 0 & 0 & 0 & 1 \end{bmatrix} \begin{bmatrix} i_d \\ i_q \\ i_o \\ i_f \end{bmatrix} \tag{7.32}$$

where:

$$\theta = \omega t$$

The stator current is

$$i_a = i_d \cos\omega t - i_q \sin\omega t \tag{7.33}$$

$$i_a = i_d \cos\omega t + i_q \cos(\omega t + 90) \tag{7.34}$$

The two components of the stator current in Equation 7.34 are shown in Figure 7.3. Note that i_d and i_q are the peak values of their own waveforms. Hence, the root mean square (rms) of these currents is

$$I_d = \frac{i_d}{\sqrt{2}}$$

$$I_q = \frac{i_q}{\sqrt{2}} \tag{7.35}$$

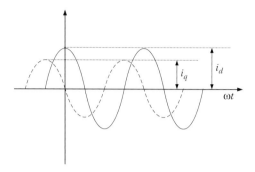

FIGURE 7.3
Waveforms of the direct and quadrature axes currents.

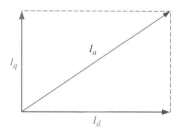

FIGURE 7.4
Phasor representation of a stator current as function of d and q currents.

In phasor form, the current in phase a is

$$\overline{I}_a = \overline{I}_d + \overline{I}_q = I_d + jI_q \tag{7.36}$$

Figure 7.4 is a phasor diagram representing Equation 7.36. The magnitude of the current in phase a is

$$I_a = \sqrt{(I_d)^2 + (I_q)^2} \tag{7.37}$$

Similarly, we can develop a phasor relationship for the voltage as

$$\overline{V}_a = \overline{V}_d + \overline{V}_q = V_d + jV_q \tag{7.38}$$

For the field winding, define the following variables:

$$V_f \equiv \frac{v_f}{\sqrt{2}}$$

$$I_f \equiv \frac{i_f}{\sqrt{2}} \tag{7.39}$$

Equation 7.30 can then be rewritten in rms form as

$$
\begin{bmatrix} V_d \\ V_q \\ V_f \end{bmatrix} = \begin{bmatrix} r_a & 0 & 0 \\ 0 & r_a & 0 \\ 0 & 0 & r_f \end{bmatrix} \begin{bmatrix} I_d \\ I_q \\ I_f \end{bmatrix} + \begin{bmatrix} 0 & -\omega L_q & 0 \\ \omega L_d & 0 & \omega L_{af} \\ 0 & 0 & 0 \end{bmatrix} \begin{bmatrix} I_d \\ I_q \\ I_f \end{bmatrix}
\tag{7.40}
$$

Equation 7.40 is known as "Parks equations," which can be rewritten as

$$
V_d = r_a I_d - x_q I_q
$$
$$
V_q = r_a I_q + x_d I_d + E_f
\tag{7.41}
$$
$$
V_f = r_f I_f
$$

where:

$$
E_f = \omega L_{af} I_f
$$
$$
x_d = \omega L_d
\tag{7.42}
$$
$$
x_q = \omega L_q
$$

where:
E_f is called *equivalent field voltage*
x_d is called *direct axis reactance*
x_q is called *quadrature axis reactance*

Because of the geometry of the airgap, the direct axis reactance is larger than the quadrature axis reactance. Equation 7.41 shows that E_f is located along the quadrature axis. In general phasor quantities, the d and q voltages in Equation 7.41 can be written as

$$
\bar{V}_d = r_a \bar{I}_d + j x_q \bar{I}_q
$$
$$
\bar{V}_q = r_a \bar{I}_q + j x_d \bar{I}_d + \bar{E}_f
\tag{7.43}
$$

The terminal voltage of the generator is then

$$
\bar{V}_a = \bar{V}_d + \bar{V}_q = r_a(\bar{I}_d + \bar{I}_q) + j x_q \bar{I}_q + j x_d \bar{I}_d + \bar{E}_f
$$
$$
\bar{V}_a = r_a \bar{I}_a + j x_q \bar{I}_q + j x_d \bar{I}_d + \bar{E}_f
\tag{7.44}
$$

For the generator operation, the currents are leaving the stator winding; hence, Equation 7.44 can be modified to

$$
\bar{E}_f = \bar{V}_a + r_a \bar{I}_a + j x_q \bar{I}_q + j x_d \bar{I}_d
\tag{7.45}
$$

The phasor diagram representing Equation 7.45 is shown in Figure 7.5. The direct axis is taken as the reference.

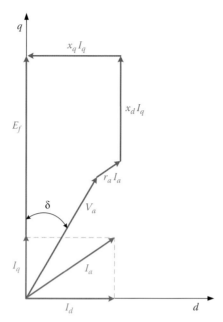

FIGURE 7.5
Phasor diagram of a salient pole synchronous generator.

EXAMPLE 7.1

A synchronous generator has the following parameters:

$$x_d = 1.0 \text{ pu} \text{ and } x_q = 0.7 \text{ pu}$$

At the rated terminal voltage, the generator is delivering rated apparent power at 0.8 lagging power factor. Compute the equivalent field voltage.

Solution:
The phasor diagram in Figure 7.5 is plotted in Figure 7.6 without the stator resistance. To locate the d and q axes, a fictitious quantity $\bar{x}_q \bar{I}_a$ as well as $\bar{E}'$ are plotted. Because $x_q < x_d \; x_q I_d < x_d I_d$.

Let us first compute the armature current

$$I_a = \frac{S_a}{V_a} = \frac{1.0}{1.0} = 1.0 \text{ pu}$$

The power factor angle is

$$\theta = \cos^{-1} 0.8 = 36.9°$$

Because the current is lagging the terminal voltage,

$$\bar{I}_a = 1.0 \times (\cos 36.9° - j \sin 36.9°) = 0.8 - j0.6 \text{ pu}$$

Take $\bar{V}_a$ as a reference and calculate E'

$$\bar{E}' = \bar{V}_a + \bar{x}_q \bar{I}_a = 1.0 + j0.7 \times (0.8 - j0.6) = 1.526 \angle 21.52° \text{ pu}$$

Hence,

$$\delta = 21.52°$$

The direct and quadrature currents are as follows:

$$I_d = I_a \, \sin(\delta + \theta) = 1.0 \times \sin(21.52° + 36.9°) = 0.852 \text{ pu}$$

$$I_q = I_a \, \cos(\delta + \theta) = 1.0 \times \cos(21.52° + 36.9°) = 0.524 \text{ pu}$$

From the phasor diagram,

$$E_f = E' + (x_d - x_q)I_d = 1.526 + (1.0 - 0.7) \times 0.852 = 1.78 \text{ pu}$$

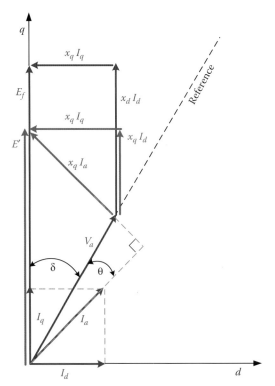

FIGURE 7.6
Phasor diagram of a salient pole generator ignoring armature losses.

7.2.3.2 Real and Reactive Powers

The complex power of the machine is the multiplication of all voltages by their corresponding conjugate current. As stated in Chapter 6, the conjugate is used to make the reactive power of the inductor positive and that for the capacitor negative. This is the norm used by the power community. The complex power of phase a in the stator is

$$\bar{S}_a = \bar{V}_a \bar{I}_a^* \tag{7.46}$$

For the three phases in the stator, the power is

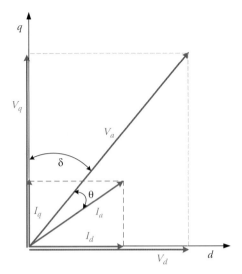

FIGURE 7.7
Direct and quadrature components of the terminal voltage and armature current.

$$\overline{S} = 3\overline{S}_a = 3\overline{V}_a\overline{I}_a^*$$ (7.47)

Substituting the current and voltage in the *dq* frames into Equation 7.47 yields

$$\overline{S} = 3\left(V_d + jV_q\right)\left(I_d + jI_q\right)^* = 3\left(V_d + jV_q\right)\left(I_d - jI_q\right)$$ (7.48)

The *d* and *q* quantities of the voltages and current can be obtained using the phasor diagram in Figure 7.7. In this case,

$$
\begin{aligned}
V_d &= V_a \sin\delta \\
V_q &= V_a \cos\delta \\
I_d &= V_a \sin(\delta + \theta) \\
I_q &= V_a \cos(\delta + \theta)
\end{aligned}
$$ (7.49)

The real power is the real component of $\overline{S}$

$$P = 3\left(V_d I_d + V_q I_q\right)$$ (7.50)

the reactive power is the imaginary component of $\overline{S}$.

$$Q = 3\left(V_q I_d - V_d I_q\right)$$ (7.51)

Substituting the value of V_d and V_q in Equation 7.49 into Equation 7.50 yields

$$P = 3\left[I_d\left(V_a \sin\delta\right) + I_q\left(V_a \cos\delta\right)\right]$$ (7.52)

For large generators, the armature resistance is small and can be ignored. In this case, as shown in Figure 7.6, the d and q currents are

$$I_d = \frac{E_f - V_a \cos\delta}{x_d}$$

$$I_q = \frac{V_a \sin\delta}{x_q}$$

(7.53)

Substituting the currents in Equation 7.53 into Equation 7.52 yields

$$P = 3\left[\frac{E_f - V_a \cos\delta}{x_d}(V_a \sin\delta) + \frac{V_a \sin\delta}{x_q}(V_a \cos\delta)\right]$$

$$= 3\frac{V_a}{x_d}\left(E_f \sin\delta + V_a \frac{x_d - x_q}{x_q} \sin\delta\cos\delta\right)$$

$$= 3\frac{V_a}{x_d}\left(E_f \sin\delta + V_a \frac{x_d - x_q}{2x_q} \sin 2\delta\right)$$

$$= P_c + P_s$$

(7.54)

where:
  P_c and P_s are known as the cylindrical and salient components of the power, respectively

$$P_c = 3\frac{E_f V_a}{x_d} \sin\delta$$

$$P_s = 3V_a^2 \frac{x_d - x_q}{2x_d x_q} \sin 2\delta$$

(7.55)

where:
  δ is known as the "power angle"

The two components of the power as well as the total power are shown in Figure 7.8.
  Similar calculations can be made for the reactive power, Equation 7.51 can be written as

$$Q = 3\left[I_d(V_a \cos\delta) - I_q(V_a \sin\delta)\right]$$

(7.56)

Substituting the current in Equation 7.53 into Equation 7.56 yields

$$Q = 3\left[\frac{E_f - V_a \cos\delta}{x_d}(V_a \cos\delta) - \frac{V_a \sin\delta}{x_q}(V_a \sin\delta)\right]$$

$$= 3V_a\left[\frac{E_f}{x_d}\cos\delta - \frac{V_a}{x_d}\cos^2\delta - \frac{V_a}{x_q}\sin^2\delta\right]$$

(7.57)

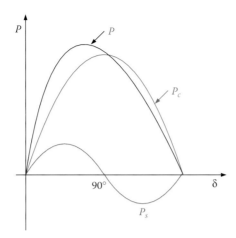

FIGURE 7.8
Power components of a salient pole synchronous generator.

EXAMPLE 7.2

For the machine in Example 7.1, if the mechanical power input to the generator increased to 1.0 per unit (pu) and the equivalent field voltage is decreased by 20%, compute the reactive power delivered to the grid.

Solution:
The increase in the mechanical power leads to an increase in the electrical power. If we examine Equation 7.54, we notice that the increase in electrical power and the change in the equivalent field voltage change the power angle. Rewriting the equation in pu quantities, we get

$$P = \frac{E_f V_a}{x_d} \sin\delta + V_a^2 \frac{x_d - x_q}{2 x_d x_q} \sin 2\delta$$

$$1.0 = \frac{(0.8 \times 1.78) \times 1.0}{1.0} \sin\delta + 1.0 \frac{1.0 - 0.7}{2 \times 1.0 \times 0.7} \sin 2\delta$$

Solving the equation numerically leads to

$$\delta = 33.87°$$

The reactive power can now be computed using Equation 7.57

$$Q = V_a \left[\frac{E_f}{x_d} \cos\delta - \frac{V_a}{x_d} \cos^2\delta - \frac{V_a}{x_q} \sin^2\delta \right]$$

$$Q = 1.0 \left[\frac{(0.8 \times 1.78)}{1.0} \cos(33.87°) - \frac{1.0}{1.0} \cos^2(33.87°) - \frac{1.0}{0.7} \sin^2(33.87°) \right] = 0.0492 \text{ pu}$$

7.3 Cylindrical Rotor Synchronous Generator

Cylindrical rotor synchronous generator is common in wind applications. The rotor of this machine is cylindrical and the airgap length is, therefore, constant. Hence, there is no change in the reluctance between the stator and any point in the rotor. This makes the direct and quadrature reactive reactances equal as seen in Equations 7.6 and 7.21.

$$x_d = x_q = x_s \tag{7.58}$$

where:
x_s is the synchronous reactance

In this case, Equation 7.45 can be written as

$$\bar{E}_f = \bar{V}_a + r_a \bar{I}_a + jx_s \bar{I}_q + jx_s \bar{I}_d$$
$$\bar{E}_f = \bar{V}_a + r_a \bar{I}_a + jx_s \bar{I}_a \tag{7.59}$$

The phasor diagram of the cylindrical rotor generator is shown in Figure 7.9. Keep in mind that E_f is always located along the quadrature axis.

The power expression in Equation 7.54 can be modified for the cylindrical rotor as

$$P = P_c = 3 \frac{E_f V_a}{x_s} \sin\delta \tag{7.60}$$

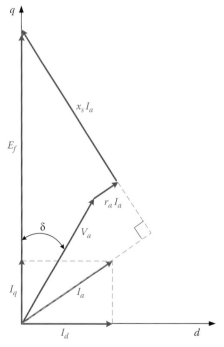

FIGURE 7.9
Phasor diagram of a cylindrical rotor synchronous generator.

Similarly, the reactive power in Equation 7.57 can be modified for the cylindrical rotor machine.

$$Q = 3\frac{V_a}{x_s}\left(E_f \cos\delta - V_a\right) \tag{7.61}$$

EXAMPLE 7.3

A synchronous generator has a synchronous reactance of 1.0 pu. At rated terminal voltage, compute the excitation voltage that delivers 1.0 pu of real power and no reactive power at the terminals of the generator.

Solution:
From Equation 7.60, we get

$$E_f \sin\delta = P\frac{x_s}{V_a}$$

$$E_f \sin\delta = 1.0$$

Because the reactive power delivered is zero, Equation 7.61 leads to

$$E_f \cos\delta = V_a = 1.0$$

Hence,

$$\delta = \tan^{-1}\frac{E_f \sin\delta}{E_f \cos\delta} = 45°$$

$$E_f = \frac{1.0}{\cos\delta} = 1.414 \text{ pu}$$

EXAMPLE 7.4

For the machine in Example 7.3, when the mechanical power input to the generator is 1.0 pu, the equivalent field voltage decreased by 20%, compute the reactive power delivered to the grid.

Solution:
If we ignore the electrical and rotational losses, the mechanical power is the same as the output electrical power. Use Equation 7.60

$$P = \frac{E_f V_a}{x_s}\sin\delta$$

$$1.0 = \frac{(0.8 \times 1.414) \times 1.0}{1.0}\sin\delta$$

$$\delta = 62.13°$$

The reactive power can now be computed using Equation 7.61

$$Q = 3\frac{V_a}{x_s}\left[E_f\cos\delta - V_a\right]$$

$$Q = 1.0\left[\frac{(0.8\times1.414)}{1.0}\cos(62.13°) - 1\right] = -0.471 \text{ pu}$$

7.4 Dynamic Model of Synchronous Generator

The dynamic model allows us to examine the behavior of the machine during disturbances. This information is essential for the evaluation of system stability as well as the interactions between the generator and the power system. The dynamic analysis is also crucial for designing protection schemes and control systems for the wind turbines. In our analysis, we are concentrating on modeling the generator when small disturbances occur, which do not activate the nonlinear modes of the system. The model is developed for the electromechanical modes of the generator. Thus, it is suitable for low frequency oscillations.

7.4.1 Dynamics of Rotating Mass

As given in Chapter 6, the dynamic equation of the rotating mass of the turbine can be written as

$$\tilde{T}_m - \tilde{T}_e = 2H\frac{d\tilde{\omega}}{dt} \tag{7.62}$$

where:
$\tilde{T}_m$ is the mechanical torque from the blade in physical units (Nm)
$\tilde{T}_e$ is the electric torque developed by the generator in physical units (Nm)
H is the inertia constant of the rotating mass (blades, generator, gearbox, etc.) in physical units (Nms²)
$\tilde{\omega}$ is the speed of the rotor in physical units (rad/s)
$d\tilde{\omega}/dt$ is the acceleration of the generator speed in rad/s²

During steady-state operation, both torques are equal. Thus, the acceleration is zero and the machine operates at constant speed. If the electrical torque becomes less than the mechanical torque, the generator speeds up, and vice versa.

To account for the disturbance in either torque, we can examine the derivative of Equation 7.62

$$\frac{d\tilde{T}_m}{dt} - \frac{d\tilde{T}_e}{dt} = 2H\frac{d^2\tilde{\omega}}{dt^2} \tag{7.63}$$

If we assume that the disturbance is small, we can use the linearization technique, whereby

$$\frac{d\tilde{T}_m}{dt} = \frac{\Delta\tilde{T}_m}{\Delta t} \tag{7.64}$$

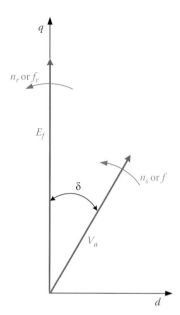

FIGURE 7.10
Power angle as a function of frequencies.

Hence, Equation 7.63 can be written as

$$\frac{\Delta \tilde{T}_m}{\Delta t} - \frac{\Delta \tilde{T}_e}{\Delta t} = 2H \frac{d}{dt} \frac{\Delta \tilde{\omega}}{\Delta t}$$

$$\Delta \tilde{T}_m - \Delta \tilde{T}_e = 2H \frac{d\Delta \tilde{\omega}}{dt} \tag{7.65}$$

Consider the phasor diagram in Figure 7.10. The diagram shows only E_f and V_a. These two phasors rotate at their own frequencies (or speeds). If the generator is connected to an infinite bus, V_a rotates at the synchronous speed (or grid frequency f). E_f rotates according to the speed of the rotor (or f_r), which is dependent on the turbine speed. If the two speeds are equal, the power angle δ is constant. However, if the turbine causes the generator to spin at a different speed than the synchronous speed, the power angle δ will change. The relationship between the change in speed and the change in δ is

$$\Delta \tilde{\omega} = \frac{d\Delta \delta}{dt} \tag{7.66}$$

In our analysis, we shall use the pu system. In this case, the pu speed is

$$\Delta \omega = \frac{\Delta \tilde{\omega}}{\omega_b} \tag{7.67}$$

where:
   ω_b is the selected base speed. Normally, it is the angular speed of the supply voltage
      (synchronous speed)

$$\Delta\omega = \frac{\Delta\tilde{\omega}}{2\pi f} \tag{7.68}$$

where:
  f is the supply frequency in Hz
  ω is the rotor speed in pu

Equation 7.66 can be rewritten as

$$\Delta\omega = \frac{1}{2\pi f}\frac{d\Delta\delta}{dt} \tag{7.69}$$

Equation 7.65 can be rewritten in pu if we divide all terms by a selected base value for the torque T_b.

$$\frac{\Delta\tilde{T}_m}{T_b} - \frac{\Delta\tilde{T}_e}{T_b} = \frac{2H}{T_b}(2\pi f)\frac{d\Delta\omega}{dt} \tag{7.70}$$

or

$$\Delta T_m - \Delta T_e = M\frac{d\Delta\omega}{dt} \tag{7.71}$$

where:
  $M = 4\pi f H / T_b$ is the equivalent inertia of the system
  ΔT_m is the change in the mechanical torque in pu
  ΔT_e is the change in the electrical torque in pu
  $\Delta\omega$ is the change in rotor speed in pu

Equations 7.69 and 7.71 are in time domain. However, frequency-domain models are often preferred for control design and stability analysis. These equations can be written in frequency domain by using Laplace transformation.

$$\Delta\omega = \frac{1}{2\pi f}S\Delta\delta \tag{7.72}$$

$$\Delta T_m - \Delta T_e = MS\Delta\omega \tag{7.73}$$

where:
  S is the Laplace operator

The above equations can be represented by the block diagram in Figure 7.11

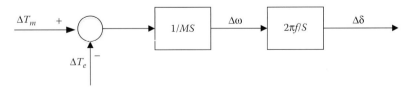

FIGURE 7.11
Model of a rotating mass.

7.4.2 Dynamics of Electrical Modes

Using Equation 7.29, we can write the three basic voltages.

$$v_d = r_a i_d - \omega \lambda_q \tag{7.74}$$

$$v_q = r_a i_q + \omega \lambda_d \tag{7.75}$$

For changing field, Equation 7.28, the field voltage is

$$v_f = r_f i_f + \frac{d\lambda_f}{dt} \tag{7.76}$$

The flux linkages in Equation 7.22 are developed for currents entering the stator. For generator, the current is leaving the stator winding; hence, we can modify the equations using a negative sign in front of the stator currents.

$$\lambda_d = -L_d\, i_d + L_{af}\, i_f \tag{7.77}$$

$$\lambda_q = -L_q\, i_q \tag{7.78}$$

$$\lambda_f = L_{ff}\, i_f - 1.5 L_{af}\, i_d \tag{7.79}$$

If we ignore the resistances of the windings for large generators, the voltages equations can be rewritten as

$$v_d = \omega L_q\, i_q \tag{7.80}$$

$$v_q = \omega L_{af}\, i_f - \omega L_d\, i_d \tag{7.81}$$

Let

$$
\begin{aligned}
x_d &= \omega L_d \\
x_q &= \omega L_q \\
x_{af} &= \omega L_{af} \\
e_f &= \omega L_{af}\, i_f
\end{aligned}
\tag{7.82}
$$

Hence,

$$v_d = x_q\, i_q \tag{7.83}$$

$$v_q = e_f - x_d\, i_d \tag{7.84}$$

The field current from Equation 7.79 is

$$i_f = \frac{1}{L_{ff}} \lambda_f + 1.5 \frac{L_{af}}{L_{ff}} i_d \tag{7.85}$$

Substituting i_f into Equation 7.77 yields

$$\lambda_d = -L_d\, i_d + L_{af}\, i_f = \frac{L_{af}}{L_{ff}} \lambda_f - \left(L_d - 1.5 \frac{L_{af}^2}{L_{ff}} \right) i_d \tag{7.86}$$

The principle of constant flux linkage states that "the flux linkage with a closed circuit having zero resistance and no voltage source cannot change." If we ignore the field resistance during a disturbance, the flux linkage λ_f is not impacted by i_d. Hence, the derivative of Equation 7.86 with respect to the current i_d is

$$-\frac{d\lambda_d}{di_d} = L_d - 1.5\frac{L_{af}^2}{L_{ff}} = L_d'$$ (7.87)

where:
  L_d' is known as the *direct axis transient reactance*

This is an important parameter for the synchronous generator. Keep in mind that L_d' is computed by ignoring the change in λ_f. With field resistance, the magnitude of λ_f changes with time as the field resistance is creating a damping effect.

7.4.2.1 Field Dynamics

If we substitute the value of i_f in Equation 7.85 into the equivalent field voltage e_f given in Equation 7.82, we get

$$e_f = \omega L_{af} i_f = \omega \frac{L_{af}}{L_{ff}}\lambda_f + 1.5\omega\frac{L_{af}^2}{L_{ff}}i_d$$ (7.88)

Rearranging the equation yields

$$e_f - \omega\left(L_d - L_d'\right)i_d = \omega\frac{L_{af}}{L_{ff}}\lambda_f = e_f'$$ (7.89)

where:
  e_f' is called the *voltage behind transient reactance*, which is on the same axis as e_f (on the quadrature axis)

Its magnitude is

$$e_f' = e_f - \left(x_d - x_d'\right)i_d$$ (7.90)

where:
  $x_d' = \omega L_d'$ is the transient reactance

The phasor diagram with the transient reactance and rms voltage behind transient reactance E_f' is shown in Figure 7.12. Note that E_f' is located along the q-axis.

To account for the dynamics in the field circuit, let us represent E_f' in terms of the filed current. From Equation 7.89, we get

$$e_f' = \omega\frac{L_{af}}{L_{ff}}\lambda_f = \omega L_{af} i_f - \omega 1.5\frac{L_{af}^2}{L_{ff}}i_d$$
$$= x_{af} i_f - 1.5\frac{x_{af}^2}{x_f}i_d$$ (7.91)

where:
  $x_f = \omega L_{ff}$
  $x_{af} = \omega L_{af}$

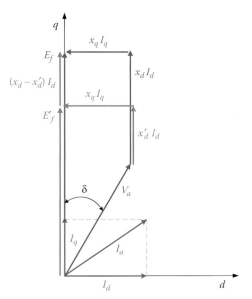

FIGURE 7.12
Phasor diagram with a transient reactance.

$$e'_f = \frac{x_{af}}{x_f}\left(x_f i_f - 1.5 x_{af}\, i_d\right) \tag{7.92}$$

Because the voltage of the field circuit is

$$v_f = r_f i_f + \frac{d\lambda_f}{dt} \tag{7.93}$$

Hence, substituting the value of the flux linkage in Equation 7.91 into Equation 7.93 yields

$$v_f = r_f i_f + \left(\frac{x_f}{\omega x_{af}}\right)\frac{de'_f}{dt} \tag{7.94}$$

$$\left(\frac{x_{af}}{r_f}\right) v_f = x_{af} i_f + \left(\frac{x_f}{\omega r_f}\right)\frac{de'_f}{dt} \tag{7.95}$$

Defining

$$e_{fd} = \left(\frac{x_{af}}{r_f}\right) v_f \tag{7.96}$$

$$\tau'_{do} = \frac{x_f}{\omega r_f} = \frac{L_{ff}}{r_f} \tag{7.97}$$

Hence,

$$e_{fd} = e_f + \tau'_{do}\frac{de'_f}{dt} \tag{7.98}$$

where:

   e_{fd} is called the *steady-state equivalent field voltage*

Equations 7.88, 7.96, and 7.98 show that $e_{fd} = e_f$ during steady state only. For constant field voltage, e_{fd} is unchanged. Substituting the value of e_f in Equation 7.90 into Equation 7.98 yields

$$e_{fd} = e'_f + \left(x_d - x'_d\right)i_d + \tau'_{do}\frac{de'_f}{dt} \tag{7.99}$$

Linearize the equation to obtain the dynamics of the field circuit.

$$\Delta e_{fd} = \Delta e'_f + \left(x_d - x'_d\right)\Delta i_d + \tau'_{do}\,S\Delta e'_f \tag{7.100}$$

7.4.2.2 Terminal Voltage Dynamics

The terminal voltage of the generator is

$$v_a^2 = v_d^2 + v_q^2 \tag{7.101}$$

In the linearized form,

$$2v_a\,\Delta v_a = 2v_d\,\Delta v_d + 2v_q\,\Delta v_q \tag{7.102}$$

$$\Delta v_a = \frac{v_d}{v_a}\,\Delta v_d + \frac{v_q}{v_a}\,\Delta v_q \tag{7.103}$$

Linearizing the values of v_d and v_q in Equations 7.80 and 7.81, we get

$$\Delta v_d = x_q\,\Delta i_q \tag{7.104}$$

$$\Delta v_q = \Delta e_f - x_d\,\Delta i_d \tag{7.105}$$

Substituting the value of e_f in Equation 7.90 into the above equation, we get

$$\Delta v_q = \left(\Delta e'_f + \left(x_d - x'_d\right)\Delta i_d\right) - x_d\,\Delta i_d \tag{7.106}$$

$$\Delta v_q = \Delta e'_f - x'_d\,\Delta i_d \tag{7.107}$$

Hence,

$$\Delta v_a = \frac{v_d}{v_a}x_q\Delta i_q + \frac{v_q}{v_a}\left(\Delta e'_f - x'_d\,\Delta i_d\right) \tag{7.108}$$

7.4.2.3 Electric Torque Dynamics

The power equation in 7.50 can be written in pu as

$$p = v_d i_d + v_q i_q \tag{7.109}$$

Linearizing the equation yields

$$\Delta p = v_d \Delta i_d + i_d \Delta v_d + v_q \Delta i_q + i_q \Delta v_q \tag{7.110}$$

Substituting Δv_d and Δv_q in Equations 7.104 and 7.105 into Equation 7.110 yields

$$\Delta p = \left(v_d - x'_d i_q \right) \Delta i_d + \left(v_q + x_q i_d \right) \Delta i_q + i_q \Delta e'_f \tag{7.111}$$

In pu system, the torque is the same as the power if the base speed is the synchronous speed of the machine. Hence,

$$\Delta T_e = \left(v_d - x'_d i_q \right) \Delta i_d + \left(v_q + x_q i_d \right) \Delta i_q + i_q \Delta e'_f \tag{7.112}$$

7.4.3 Block Diagram of Synchronous Generator

Equations 7.72, 7.73, 7.100, 7.108, and 7.112 represent the dynamics of the synchronous generator. These equations can be represented by a block diagram that allows us to analyze system response, check system stability, and design control systems.

In most of the equations mentioned above, the currents are needed. Hence, our initial step is to find these currents, which can be obtained from the phasor diagram in Figure 7.12, where

$$v_d = v_a \sin \delta = x_q \, i_q \tag{7.113}$$

$$v_q = v_a \cos \delta = e'_f - x'_d i_d \tag{7.114}$$

Hence,

$$\begin{bmatrix} i_d \\ i_q \end{bmatrix} = \begin{bmatrix} 1 \\ 0 \end{bmatrix} \frac{e'_f}{x'_d} + \begin{bmatrix} \dfrac{-\cos \delta}{x'_d} \\ \dfrac{\sin \delta}{x_q} \end{bmatrix} v_a \tag{7.115}$$

Linearizing Equation 7.115 yields

$$\Delta i_d = \frac{1}{x'_d} \Delta e'_f + F_d \Delta \delta + C_d \Delta v_a \tag{7.116}$$

$$\Delta i_q = F_q \Delta \delta + C_q \Delta v_a$$

where:

$$\begin{bmatrix} F_d \\ F_q \end{bmatrix} = \begin{bmatrix} \dfrac{\sin\delta}{x_d'} \\ \dfrac{\cos\delta}{x_q} \end{bmatrix} v_a$$

$$\begin{bmatrix} C_d \\ C_q \end{bmatrix} = \begin{bmatrix} \dfrac{-\cos\delta}{x_d'} \\ \dfrac{\sin\delta}{x_q} \end{bmatrix}$$

(7.117)

Substitute the values of Δi_d and Δi_q in Equation 7.116 into the electric torque Equation 7.112.

$$\Delta T_e = \left(v_d - x_d' i_q\right)\left(\frac{1}{x_d'}\Delta e_f' + F_d\Delta\delta + C_d\Delta v_a\right) + \left(v_q + x_q i_d\right)\left(F_q\Delta\delta + C_q\Delta v_a\right) + i_q\Delta e_f' \qquad (7.118)$$

$$\Delta T_e = \left(v_d F_d - x_d' i_q F_d + v_q F_q + x_q i_d F_q\right)\Delta\delta + \frac{v_d}{x_d'}\Delta e_f' + \left(v_d C_d - x_d' i_q C_d + v_q C_q + x_q i_d C_q\right)\Delta v_a \qquad (7.119)$$

or

$$\Delta T_e = K_1\Delta\delta + K_2\Delta e_f' + K_3\Delta v_a \qquad (7.120)$$

where:

$$K_1 = v_d F_d - x_d' i_q F_d + v_q F_q + x_q i_d F_q$$

$$K_2 = \frac{v_d}{x_d'}$$

$$K_3 = v_d C_d - x_d' i_q C_d + v_q C_q + x_q i_d C_q$$

The next equation is Δe_{fd}. Substituting the values of currents in Equation 7.116 into Equation 7.100 yields

$$\Delta e_{fd} = \Delta e_f' + \left(x_d - x_d'\right)\left(\frac{1}{x_d'}\Delta e_f' + F_d\Delta\delta + C_d\Delta v_a\right) + \tau_{do}' S\Delta e_f' \qquad (7.121)$$

$$\Delta e_f' = \frac{1}{\left[\left(x_d/x_d'\right) + \tau_{do}' S\right]}\left[\Delta e_{fd} - \left(x_d - x_d'\right)F_d\Delta\delta - \left(x_d - x_d'\right)C_d\Delta v_a\right]$$

$$= \frac{K_4}{\left(1 + \tau_{do}' K_4 S\right)}\left(\Delta e_{fd} - K_5\Delta\delta - K_6\Delta v_a\right)$$

where:

$$K_4 = \frac{x_d'}{x_d}$$

$$K_5 = \left(x_d - x_d' \right) F_d$$

$$K_6 = \left(x_d - x_d' \right) C_d$$

The last equation is the terminal voltage. Substitute the currents in Equation 7.116 into Equation 7.108.

$$\Delta v_a = \frac{v_d}{v_a} x_q \left(F_q \Delta \delta + C_q \Delta v_a \right) + \frac{v_q}{v_a} \left[\Delta e_f' - x_d' \left(\frac{1}{x_d'} \Delta e_f' + F_d \Delta \delta + C_d \Delta v_a \right) \right]$$

$$= \frac{v_d\, x_q F_q\ -\ v_q\, x_d' F_d}{v_a\ -\ v_d\, x_q C_q\ +\ v_q\, x_d' C_d} \Delta \delta \qquad\qquad (7.122)$$

$$= K_7 \Delta \delta$$

where:

$$K_7 = \frac{v_d\, x_q F_q\ -\ v_q\, x_d' F_d}{v_a\ -\ v_d\, x_q C_q\ +\ v_q\, x_d' C_d}$$

Based on Equations 7.72, 7.73, 7.120, 7.121, and 7.122, we can construct the block diagram of the synchronous generator in Figure 7.13. Keep in mind that K_1 to K_7 are computed at the steady-state condition before the disturbance.

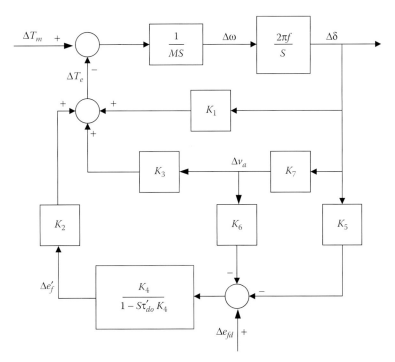

FIGURE 7.13
Block diagram of a synchronous generator.

The model in Figure 7.13 can be modified for cylindrical rotor generator and for permanent magnetic machines. For the cylindrical rotor, we use the relationship in Equation 7.58. For the permanent magnetic machine, e_{fd} is constant. Hence, $\Delta e_{fd} = 0$ in the model.

Exercise

1. What are the main components of synchronous generator?
2. What is the difference between salient and cylindrical rotor machines?
3. What determines the frequency of the output voltage of the synchronous generator?
4. What are the inductances that change with rotor position?
5. What is the equivalent field voltage?
6. A 1 GW cylindrical rotor synchronous generator is connected to a grid through a transmission line. The synchronous reactance of the generator is 9 Ω and the inductive reactance of the transmission line is 3 Ω. The grid voltage is 110 kV. If the generator delivers its rated power at unity power factor to the infinite bus, compute the following:
   a. Terminal voltage of the generator
   b. Equivalent field voltage
   c. Real power output at the generator terminals
   d. Reactive power output at the generator terminals
7. A cylindrical rotor synchronous generator is connected directly to an infinite bus. The voltage of the infinite bus is 15 kV. The excitation of the generator is adjusted until the equivalent field voltage E_f is 14 kV. The synchronous reactance of the machine is 5 Ω. Compute the following:
   a. Pullover power (maximum power)
   b. Equivalent excitation voltage that increases the pullover power by 20%
8. A type 4 wind turbine system employing cylindrical synchronous generator is connected to the grid through an ac/ac converter. The output of the generator is 800 V (line-to-line) at 30 Hz. The point of interconnection at the grid is 690 V (line-to-line). The transmission line between the converter and the grid has inductive reactance of 1.0 Ω. The ac/dc part of the converter is made of full-wave, three-phase diode-bridge circuit. The dc/ac part of the converter is made of a three-phase, full-wave switching circuit with PWD. Ignore all harmonics and compute the following:
   a. Voltage of the dc bus
   b. Amplitude modulation of the PWM that results in 700 V (line-to-line) at the output of the converter
   c. Phase angle of the voltage that delivers 400 kW of real power to the grid
   d. Reactive power delivered to the grid

9. A salient pole synchronous generator has the following parameters:

$x_d = 1.0$ pu; $x_q = 0.6$ pu; $S = 1.2$ pu at 0.8 pf lagging current; and $V_a = 1.0$ pu

Compute the excitation voltage.

10. For the system in Problem 9, compute the real and reactive power at the terminals of the generator.

11. For the system in Problem 10, compute the cylindrical and salient components of the real power.

12. For the system in Problem 11, compute the power angle at maximum real power.

8

Type 1 Wind Turbine System

Type 1 system is among the oldest designs of wind turbines. It employs the squirrel-cage induction generator, because of its inexpensive cost, low maintenance, and high reliability. Also, unlike the synchronous generator, the induction generator does not have to spin at a precise speed to deliver electricity to the grid. It is capable of producing electricity at the grid frequency even when the rotor speed varies. Thus, no conditioning device is needed between the terminals of the generator and the grid, except at starting.

The rotor windings of type 1 generator are not accessible from a stationary position. This makes type 1 wind turbine the least-controlled system, which is mainly used in small wind turbines.

8.1 Equivalent Circuit for the Squirrel-Cage Induction Generator

In Chapter 6, the equivalent circuit is developed, and it is shown here in Figure 8.1. V_D in the figure is

$$\bar{V}_D = -r_d\bar{I}'_{a2} = -\frac{r'_2}{s}(1-s)\bar{I}'_{a2} \tag{8.1}$$

For the squirrel-cage machine, there is no injection in the rotor circuit. Hence, $V'_{a2} = 0$, and the equivalent circuit in this case is shown in Figure 8.2.

8.1.1 Power Flow

The power flow of the squirrel-cage induction generator is shown in Figure 8.3. The output of the gearbox enters the generator where part of it is lost due to rotational losses (friction, windage, etc.). The rest of the power is converted into electrical power, called "developed power," P_d.

$$P_d = 3V_D I'_{a2} \cos\theta = -3\frac{r'_2}{s}(1-s)(I'_{a2})^2 \tag{8.2}$$

where:
θ is the angle between V_D and I'_{a2}

The developed power can also be computed using mechanical variables (torque and speed). The torque is called developed torque and the speed is the speed of the rotor ω_2.

$$P_d = T_d\omega_2 \tag{8.3}$$

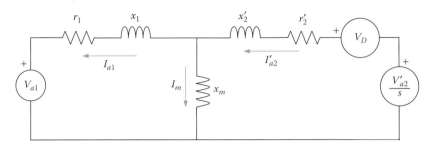

FIGURE 8.1
Generic induction generator model.

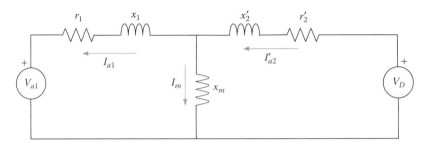

FIGURE 8.2
Model of a squirrel-cage induction generator.

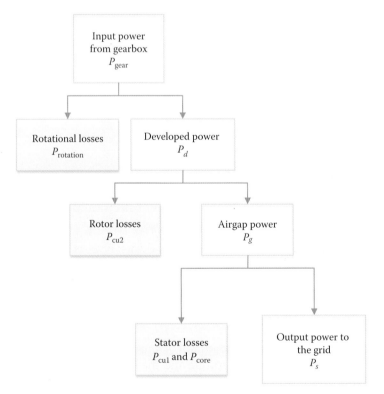

FIGURE 8.3
Power flow of a squirrel-cage induction generator.

Part of the developed power is lost in the resistance of the rotor winding in the form of copper losses P_{cu2}.

$$P_{cu2} = 3 \ r_2' \ (I_{a2}')^2 \tag{8.4}$$

The rest of the power goes to the stator through the airgap flux. Thus, it is called "airgap power," P_g.

$$P_g = P_d - P_{cu2} = -3 \frac{r_2'}{s} (I_{a2}')^2 \tag{8.5}$$

The airgap power can also be computed using the torque and speed of the airgap flux. The torque is the developed torque and the speed is the speed of the airgap flux, which is the synchronous speed, ω_s.

$$P_g = T_d \omega_s \tag{8.6}$$

Comparing Equations 8.2 and 8.5, we get

$$P_g = \frac{P_d}{(1-s)} \tag{8.7}$$

When P_g enters the stator, part of it is lost in the resistance of the stator winding in the form of copper losses P_{cu1}.

$$P_{cu1} = 3 \ r_1 \ I_{a1}^2$$

Another loss in the stator is the core loss P_{core}. The rest of the power (P_s) leaves the stator terminals and is delivered to the grid.

$$P_s = P_g - P_{cu1} - P_{core} = 3V_{a1}I_{a1} \cos\theta_1 \tag{8.8}$$

where:
θ_1 is the angle between V_{a1} and I_{a1}

EXAMPLE 8.1

An induction generator of type 1 system is running at a speed of 1230 rpm. The power delivered to the generator is 400 KW. Ignore the stator losses and compute the power delivered to the grid.

Solution:
Because type 1 system is a squirrel-cage machine, it runs at super-synchronous speed. Also, as the machine is considered fixed speed, the synchronous speed is very close to actual speed. Hence, at 60 Hz, the machine must be a six-pole type, which has a synchronous speed of 1200 rpm.

The slip of the machine at 1230 rpm is

$$s = \frac{n_s - n}{n_s} = \frac{1200 - 1230}{1200} = -0.025$$

If we ignore the stator losses, the power delivered to the grid is the airgap power.

$$P_g = \frac{P_d}{(1-s)} = \frac{400}{(1+0.025)} = 390.24 \ KW$$

To compute the voltage–current relationship, we can use the equivalent circuit in Figure 8.2. To simplify the circuit, Thevinin's theorem can be applied to the stator circuit, as shown in Figure 8.4, where Thevinin's voltage, $\bar{V}_{th}$, and Thevinin's impedance, $\bar{Z}_{th}$, are as follows:

$$\bar{V}_{th} = jx_m\bar{I} = jx_m\frac{\bar{V}_{a1}}{r_1 + j(x_1 + x_m)} \tag{8.9}$$

$$\bar{Z}_{th} = \frac{jx_m(r_1 + jx_1)}{r_1 + j(x_1 + x_m)} = r_{th} + jx_{th} \tag{8.10}$$

Next, we add the rotor circuit to the Thevinin's equivalent circuit, as shown in Figure 8.5. Further, we can lump the parameters of the induction machine to obtain the simple equivalent circuit in Figure 8.6, where

$$r_{eq} = r_{th} + r_2'$$

$$x_{eq} = x_{th} + x_2' \tag{8.11}$$

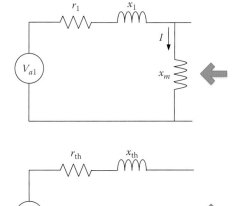

FIGURE 8.4
Thevinin's equivalent of stator.

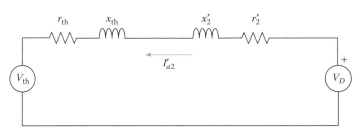

FIGURE 8.5
Modified squirrel-cage induction generator model.

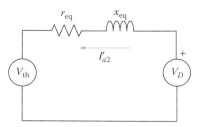

FIGURE 8.6
Model of a squirrel-cage induction generator with lumped parameters.

The rotor current can be written as

$$\bar{I}'_{a2} = \frac{\bar{V}_D - \bar{V}_{th}}{r_{eq} + jx_{eq}} \tag{8.12}$$

Substituting the value of V_D in Equation 8.1 into Equation 8.12 yields the magnitude of the rotor current.

$$I'_{a2} = \frac{-V_{th}}{\sqrt{\left[r_{th} + \left(r'_2/s \right) \right]^2 + x_{eq}^2}} \tag{8.13}$$

EXAMPLE 8.2

A type 1 wind turbine with a six-pole, 60 Hz three-phase, Y-connected induction generator is spinning the generator at 1250 r/min. The terminal voltage of the generator is 690 V. The parameters of the machine are: $r_1 = r'_2 = 10\ m\Omega$; $x_1 = x'_2 = 100\ m\Omega$; and $x_m = 2\ \Omega$. Compute the stator current.

Solution:
The slip of the machine at 1250 rpm is

$$s = \frac{n_s - n}{n_s} = \frac{1200 - 1250}{1200} = -0.0417$$

To compute the rotor current of the generator, we need to compute Thevinin's voltage and impedance. Thevinin's voltage can be computed from Equation 8.9.

$$\bar{V}_{th} = jx_m \frac{\bar{V}_{a1}}{r_1 + j(x_1 + x_m)} = j5 \frac{\left(690/\sqrt{3} \right)\angle 0°}{0.01 + j2.1} = 379.39 \angle 0.27°\,V$$

Thevinin's impedance can be computed using Equation 8.10.

$$\bar{Z}_{th} = \frac{jx_m(r_1 + jx_1)}{r_1 + j(x_1 + x_m)} = \frac{j2 \times (0.01 + j0.1)}{0.01 + j2.1} = 0.00907 + j0.09528\ \Omega$$

Now we can compute the rotor current using Equation 8.13.

$$\bar{I}'_{a2} = \frac{-\bar{V}_{th}}{\left[r_{th} + \left(r'_2/s \right) \right] + j\left(x_{th} + x'_2 \right)} = -\frac{379.39 \angle 0.27°}{\left[0.00907 + \left(0.01/-0.0417 \right) \right] + j0.19528} = 1.562 \angle 53.77°\,kA$$

The voltage across the magnetizing branch (core) V_m can be calculated using the rotor loop.

$$\overline{V}_m + \overline{I}'_{a2}\left(\frac{r'_2}{s} + jx'_2\right) = 0$$

$$\overline{V}_m = -\overline{I}'_{a2}\left(\frac{r'_2}{s} + jx'_2\right) = -1.562\ \angle 53.77° \times \left(\frac{0.01}{-0.0417} + j0.1\right) = 405.75\ \angle 31.13°\,\text{V}$$

The stator current is

$$\overline{I}_{a1} = \frac{\overline{V}_m - \overline{V}_{a1}}{r_1 + jx_1} = \frac{405.75\ \angle 31.13° - \left(690/\sqrt{3}\right)}{0.01 + j0.1} = 1.527\ \angle 58.68°\,\text{kA}$$

The characteristics of the developed power versus speed of the generator can be obtained from Equations 8.5 and 8.7.

$$P_d = P_g(1-s) = -3\frac{r'_2}{s}I'^2_{a2}(1-s) \tag{8.14}$$

Substituting the value of I'_{a2} in Equation 8.13 into Equation 8.14 yields

$$P_d = -3\frac{(1-s)}{s}\frac{r'_2 V^2_{\text{th}}}{\left[r_{\text{th}} + \left(r'_2/s\right)\right]^2 + x^2_{\text{eq}}} \tag{8.15}$$

Equation 8.15 is plotted in Figure 8.7. The figure shows that the machine generates electricity when the speed is above the synchronous speed ω_s, when the slip is negative. The figure

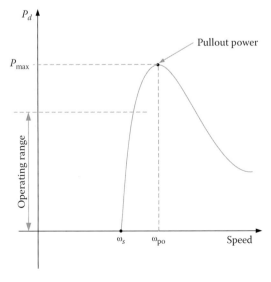

FIGURE 8.7
Power–speed characteristic of a squirrel-cage induction generator.

also shows that the slope of the curve is very steep due to the small rotor resistance of large generators. Because of the steep slope, the power from the gearbox can change widely without any appreciable change in the rotating speed. This is why type 1 system is called "fixed-speed turbine." The maximum developed power of the generator is called "pullout power" and it occurs at a speed called "pullout speed," ω_{po}. Beyond ω_{po}, the developed power of the generator decreases with the increase in speed causing the machine to speed up and eventually damaged due to over speeding. This is unstable operation as discussed later in this chapter.

The control actions for type 1 system are very limited. Most of the controls are relying on the mechanical components to provide torque and power adjustments; mostly through pitching the blades.

8.1.2 Electric Torque

The developed torque, which is also known as "electric torque," can be obtained from Equation 8.6

$$T_d = \frac{P_g}{\omega_s} = \frac{P_d}{\omega_s(1-s)} \tag{8.16}$$

Substituting P_d in Equation 8.15 into Equation 8.16 yields

$$T_d = -3\frac{r_2'}{s\omega_s}\frac{V_{th}^2}{\left[r_{th} + \left(r_2'/s\right)\right]^2 + x_{eq}^2} \tag{8.17}$$

Equation 8.17 represents the torque–speed characteristic of the generator, which is plotted in Figure 8.8. It is very similar in shape to the power–speed curve in Figure 8.7. The theoretical maximum developed torque of the generator is known as the "pullout torque" (T_{po}) or "breakdown torque." The speed at the pullout torque is called "pullout speed" (ω_{po}).

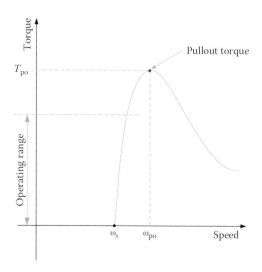

FIGURE 8.8
Torque–speed characteristic of a squirrel-cage induction generator.

EXAMPLE 8.3

A type 1 wind turbine with a six-pole, 60 Hz three-phase, Y-connected induction generator is spinning the generator at 1250 r/min. The terminal voltage of the generator is 690 V. The parameters of the machine are $r_1 = r_2' = 10\,\text{m}\Omega$; $x_1 = x_2' = 100\,\text{m}\Omega$; and $x_m = 2\,\Omega$.

Ignore the core losses and the rotational losses of the generator. Compute the following:

1. Current in the rotor circuit referred to stator
2. Developed power
3. Airgap power
4. Power delivered to the grid

Solution:

1. To compute the rotor current of the generator, we need to compute Thevinin's voltage and impedance. Thevinin's voltage can be computed from Equation 8.9.

$$\bar{V}_{th} = jx_m \frac{\bar{V}_{a1}}{r_1 + j(x_1 + x_m)} = j2 \frac{\left(690/\sqrt{3}\right)\angle 0°}{0.01 + j2.1} = 379.39\ \angle 0.27° \text{V}$$

Thevinin's impedance can be computed using Equation 8.10.

$$\bar{Z}_{th} = \frac{jx_m(r_1 + jx_1)}{r_1 + j(x_1 + x_m)} = \frac{j2 \times (0.01 + j0.1)}{0.01 + j2.1} = 0.00907 + j0.09528\ \Omega$$

For a six-pole machine, the synchronous speed is

$$n_s = 120\frac{f}{p} = 120\frac{60}{6} = 1200\,\text{r/min}$$

The slip of the machine is

$$s = \frac{n_s - n}{n_s} = \frac{1200 - 1250}{1200} = -0.0417$$

Now we can compute the rotor current using Equation 8.13

$$\bar{I}_{a2}' = \frac{-\bar{V}_{th}}{\left[r_{th} + \left(r_2'/s\right)\right] + j\left(x_{th} + x_2'\right)} = -\frac{379.39\ \angle 0.27°}{\left(0.00907 + \dfrac{0.01}{-0.0417}\right) + j0.19528} = 1.255\ \angle 40.23°\,\text{kA}$$

2. The developed power is obtained from Equation 8.2

$$P_d = -3\frac{r_2'}{s}(1-s)\left(I_{a2}'\right)^2 = -3\frac{0.01}{-0.0417}(1+0.0417)(1255)^2 = 1.18\,\text{MW}$$

3. The airgap power is

$$P_g = P_d - P_{cu2} = P_d - 3r_2'\left(I_{a2}'\right)^2 = 1.18 - 3 \times 0.01(1255)^2 \times 10^{-6} = 1.132\,\text{MW}$$

4. Next step is to compute the stator current. But first, compute the voltage $\bar{V}_m$ across the magnetizing branch using the rotor loop.

$$\bar{V}_m + \bar{I}'_{a2}\left(\frac{r'_2}{s} + jx'_2\right) = 0$$

$$\bar{V}_m = -\bar{I}'_{a2}\left(\frac{r'_2}{s} + jx'_2\right) = -1.255\,\angle 40.23° \times \left(\frac{0.01}{-0.0417} + j0.1\right) = 326.08\,\angle 17.59°\,\text{V}$$

The stator current is

$$\bar{I}_{a1} = \frac{\bar{V}_m - \bar{V}_{a1}}{r_1 + jx_1} = \frac{326.08\,\angle 17.59° - \left(690/\sqrt{3}\right)}{0.01 + j0.1} = 1.312\,\angle 47.31°\,\text{kA}$$

5. The power delivered to the grid is then

$$P_s = P_g - 3r_1\left(I'_{a1}\right)^2 = 1.132 - 3 \times 0.01 \times (1255)^2 \times 10^{-6} = 1.085\,\text{MW}$$

EXAMPLE 8.4

A wind turbine has wind blades of 25 m long. The far stream wind speed is 15 m/s and the coefficient of performance C_p is 30% at a given pitch angle. The induction generator is a six-pole machine and is rotating at 1260 r/min. The efficiency from the blades to the generator is 85% and the efficiency of the generator is 90%. Compute the following:

1. Power captured by the blade
2. Slip of the generator
3. Power delivered to the stator
4. Output electric power of the wind turbine
5. Rotor copper loss
6. Stator losses
7. Total system efficiency

Solution:
1. Compute the power in the upwind, as shown in Chapter 2

$$P_w = \frac{1}{2}\delta A_{\text{blade}}\,w_u^3 = \frac{1}{2}(\pi \times 25^2) \times 15^3 = 3.31\,\text{MW}$$

The blade power is

$$P_{\text{blade}} = P_w C_p = 3.31 \times 0.3 = 993\,\text{KW}$$

2. To compute the slip, we need to compute the synchronous speed of the machine first.

$$n_s = 120\frac{f}{p} = 120\frac{60}{6} = 1200\,\text{rpm}$$

The slip is then

$$s = \frac{n_s - n}{n_s} = \frac{1200 - 1260}{1200} = -0.05$$

3. The power delivered to the stator is the airgap power. But first, we need to compute the developed power P_d, which can be computed from the blade power P_{blade}

$$P_d = P_{\text{blade}}\,\eta_m = 993 \times 0.85 = 844.05\,\text{kW}$$

where:
   η_m is the mechanical efficiency of the turbine; it account for the rotational losses (friction, windage, etc.)

The airgap power of the generator P_g is related to the developed by

$$P_g = \frac{P_d}{(1-s)} = \frac{844.05}{1-0.05} = 803.86\,\text{kW}$$

4. The output power of the generator is

$$P_{\text{out}} = P_s = P_d\,\eta_e = 844.05 \times 0.9 = 759.65\,\text{kW}$$

where:
   η_e is the electrical efficiency of the generator

5. Rotor copper loss is

$$P_{\text{cu2}} = P_d - P_g = 844.05 - 803.86 = 40.19\,\text{kW}$$

6. Stator losses is

$$P_{\text{loss1}} = P_{\text{cu1}} + P_c = P_g - P_{\text{out}} = 8403.86 - 759.65 = 44.21\,\text{kW}$$

7. Total system efficiency is

$$\eta_{\text{total}} = C_p\,\eta_m\,\eta_e = 0.3 \times 0.85 \times 0.9 = 22.95\%$$

8.1.3 Maximum Power

As seen in Figures 8.7 and 8.8, any induction generator has maximum values for electric power (developed power) and electric torque (developed torque). If the mechanical power or torque from the gearbox exceeds these values, the machine will speed up and may be damaged. This can be seen from the dynamic electromechanical equation of rotating mass

$$T_m - T_d = \Delta T = J\frac{d\omega}{dt} \tag{8.18}$$

where:

T_d is the developed electrical torque of the generator
T_m is the mechanical torque from the gearbox
J is the inertia of the system
$d\omega/dt$ is the acceleration of rotor speed

Consider the torque–speed characteristics in Figure 8.8. If the mechanical torque is less than the maximum value (pullout torque), the machine operates at a stable speed where the two torques are equal. The speed in this case is any value between ω_s and ω_{po}. Let us assume that the mechanical torque from the gearbox increases to the pullout torque, in this case, the machine, theoretically, still operating at a constant speed ω_{po}. If the mechanical torque increases beyond the pullout torque, the difference in the two torques causes an acceleration that increases the speed of the generator to beyond ω_{po} as given in Equation 8.18. When the speed is higher than ω_{po}, the following sequence occurs:

1. Electric torque decreases, as seen in Figure 8.8.
2. ΔT in Equation 8.18 increases.
3. Generator accelerates even faster.
4. If mechanical torque is not reduced quickly, the machine can spin out of control and the turbine can be damaged.

The above scenario represents unstable operation. To avoid this problem, the generator must always operate below its pullout torque and pullout power.

To compute the pullout power and the slip at the pullout power, we can write Equations 8.5 and 8.7 as

$$P_d = P_g(1-s) = -3\frac{r_2'}{s}I_{a2}'^2(1-s) \tag{8.19}$$

Substituting the value of I_{a2}' in Equation 8.13 into Equation 8.19 yields

$$P_d = -3\frac{(1-s)}{s}\frac{r_2'V_{th}^2}{\left[r_{th}+\left(r_2'/s\right)\right]^2+x_{eq}^2} \tag{8.20}$$

To compute the pullout (maximum) developed power, we equate the derivative of Equation 8.20 to zero with respect to the slip.

$$\frac{dP_d}{ds} = -\frac{d}{ds}\left\{3\frac{(1-s)}{s}\frac{r_2'V_{th}^2}{\left[r_{th}+\left(r_2'/s\right)\right]^2+x_{eq}^2}\right\} = 0 \tag{8.21}$$

The solution leads to the slip at pullout power (s'), which is

$$s' = -m\left[m+\sqrt{m^2+1}\right] \tag{8.22}$$

where:

$$m = \frac{r_2'}{\sqrt{r_{th}^2+2r_{th}r_2'+x_{eq}^2}}$$

Substituting Equation 8.22 into Equation 8.20 yields the pullout power (maximum power).

$$P_{d-\max} = -3 \frac{(1-s')}{s'} \frac{r_2' V_{\text{th}}^2}{\left[r_{\text{th}} + \left(r_2'/s' \right) \right]^2 + x_{\text{eq}}^2} \tag{8.23}$$

EXAMPLE 8.5

A type 1 wind turbine has a six-pole, 60 Hz three-phase, Y-connected induction generator. The terminal voltage of the generator is 690 V. The rotor parameters of the machine are $r_2' = 10\,\text{m}\Omega$ and $x_2' = 100\,\text{m}\Omega$.

Assume the impedance of the stator circuit is much smaller than the impedance of the magnetizing branch. Compute the pullout power.

Solution:
If we ignore the stator parameters, the following approximation can be made:

$$V_{\text{th}} \approx V_{a1}$$

$$r_{\text{th}} \approx 0$$

$$x_{\text{eq}} \approx x_2'$$

To compute the slip at pullout power, we need to compute m in Equation 8.22.

$$m = \frac{r_2'}{\sqrt{r_{\text{th}}^2 + 2 r_{\text{th}} r_2' + x_{\text{eq}}^2}} \approx \frac{r_2'}{x_2'} = 0.1$$

$$s' = -m \left[m + \sqrt{m^2 + 1} \right] = -0.1 \times \left[0.1 + \sqrt{0.1^2 + 1} \right] = -0.1105$$

The speed at the pullout power is

$$n' = n_s(1 - s') = 1200 \times (1 + 0.1105) = 1332.6\,\text{rpm}$$

The change in speed to pullout power is

$$\Delta n' = n_s - n' = 1200 - 1332.6 = 132.6\,\text{rpm}$$

The pullout power (maximum power) is

$$P_{d-\max} = -3 \frac{(1-s')}{s'} \frac{V_{a1}^2 r_2'}{\left(r_2'/s' \right)^2 + \left(x_2' \right)^2} = 3 \frac{(1+0.1105)}{0.1105} \frac{(690^2/3)0.01}{(0.01/-0.1105)^2 + (0.1)^2} = 2.53\,\text{MW}$$

8.1.4 Maximum Torque

Similar process can be made for the maximum developed torque. The developed torque equation is

$$T_d = \frac{P_d}{\omega_2} = \frac{P_d}{\omega_s(1-s)} = -\frac{3}{s\omega_s} \frac{r_2' V_{\text{th}}^2}{\left[r_{\text{th}} + \left(r_2'/s \right) \right]^2 + x_{\text{eq}}^2} \tag{8.24}$$

The maximum torque is obtained by equating the derivative of the torque equation to zero with respect to the slip.

$$\frac{dT_d}{ds} = -\frac{d}{ds}\left\{\frac{3}{s\omega_s}\frac{r_2'V_{th}^2}{\left[r_{th}+\left(r_2'/s\right)\right]^2+x_{eq}^2}\right\} = 0 \tag{8.25}$$

where the solution of Equation 8.25 leads to the slip at maximum torque s^*

$$s^* = -\frac{r_2'}{\sqrt{r_{th}^2+x_{eq}^2}} \tag{8.26}$$

The negative sign in Equation 8.26 is for generators at super-synchronous speed. Substituting the value of s^* in Equation 8.26 into Equation 8.24 yields

$$T_{d-max} = -\frac{3}{s^*\omega_s}\frac{r_2'V_{th}^2}{\left[r_{th}+\left(r_2'/s^*\right)\right]^2+x_{eq}^2} = \frac{3V_{th}^2}{2\omega_s\left(\sqrt{r_{th}^2+x_{eq}^2}-r_{th}\right)} \tag{8.27}$$

EXAMPLE 8.6

For the system in Example 8.5, compute the maximum torque and the speed at maximum torque.

Solution:
The slip at maximum torque is

$$s^* = -\frac{r_2'}{x_2'} = \frac{0.01}{0.1} = -0.1$$

The speed at the maximum torque is

$$n^* = n_s\left(1-s^*\right) = 1200\times(1+0.1) = 1320\,\text{rpm}$$

The maximum speed range is

$$\Delta n^* = n_s - n^* = 1200 - 1320 = 120\,\text{rpm}$$

The maximum torque is

$$T_{d-max} = \frac{3}{2}\frac{V_{a1}^2}{x_2'\omega_s} = \frac{690^2}{2\times0.1\times40\pi} = 18.94\,\text{kNm}$$

8.2 Assessment of Type 1 System

Type 1 system employs the squirrel-cage induction generator. Although inexpensive and rugged, the squirrel-cage generator has some drawbacks. The following are some of the drawbacks:

- Without the rotating flux in the airgap, the generator cannot produce any power. Because the flux is formed by the grid voltage, the generator must be connected to the grid. Hence, type 1 turbine cannot operate as a stand-alone system.
- Squirrel-cage induction generator consumes reactive power from the grid as it cannot generate its own. Thus, type 1 system cannot provide voltage support to the grid.
- The generator draws high currents at starting. This can cause a dip in voltage and may cause the machine to trip due to excessive over currents.
- During voltage suppression, the generator can speed up out of control.

8.3 Control and Protection of Type 1 System

To address some of the drawbacks of type 1 system, and to prevent the turbine from frequent tripping or even damage, several control and protection equipment must be used. In the next subsections, we address the reactive power, inrush current, and the electromechanical stress. The basic control and protection is given in each subsection. In Chapter 12, more methods are discussed in more details.

8.3.1 Reactive Power of Type 1 System

The inductive elements of the induction generator consumes substantial amount of reactive power that cannot be produced by the generator itself as it lacks its own excitation. This reactive power must come from external source such as the grid. The reactive power is consumed by three main components: Q_1, which is consumed by the inductive reactance of the stator windings x_1; Q_2, which is consumed by the inductive reactance of the rotor windings x_2'; and Q_m, which is consumed by the inductive reactance of the core x_m.

$$Q_1 = 3\, x_1 I_{a1}^2 \tag{8.28}$$

$$Q_2 = 3 x_2' \left(I_{a2}'\right)^2 \tag{8.29}$$

$$Q_m = 3\, x_m I_m^2 \tag{8.30}$$

We can add these three reactive powers to compute the total consumption Q.

$$Q = Q_1 + Q_2 + Q_m \tag{8.31}$$

Q_1, Q_2, and, to some extent, Q_m depend on the rotor current, which is a function of the speed of the generator as given in Equation 8.13. Figure 8.9 shows the reactive power consumed by the machine as a function of speed. Within the operating range of speed, the reactive power is almost linearly proportional to the speed. Thus, fluctuations in wind speed causes fluctuation in reactive power, which leads to voltage fluctuations in weak grids.

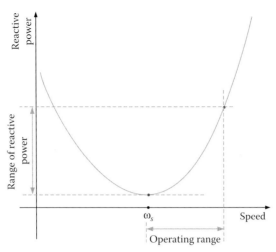

FIGURE 8.9
Reactive power demand of an induction machine.

EXAMPLE 8.7

For the system in Example 8.3, compute the reactive power of the generator and the power factor at the terminals of the generator.

Solution:
In Example 8.3, we computed the following:

$$\bar{I}'_{a2} = 1.562 \ \angle 53.77° \,\text{kA}$$

$$\bar{V}_m = 405.75 \ \angle 31.13° \,\text{V}$$

$$\bar{I}_{a1} = 1.527 \ \angle 58.68° \,\text{kA}$$

The current in the core inductive reactance is

$$\bar{I}_m = \frac{\bar{V}_m}{jx_m} = 202.88 \ \angle -58.87° \,\text{A}$$

The reactive power in the stator winding is

$$Q_1 = 3x_1 I_{a1}^2 = 3 \times 0.1 \times 1.527^2 = 700 \,\text{kVAr}$$

The reactive power of the core is

$$Q_m = 3x_m I_m^2 = 3 \times 2 \times 202.88^2 = 246.9 \,\text{kVAr}$$

The reactive power of the rotor windings

$$Q_2 = 3x'_2 \left(I'_{a2} \right)^2 = 3 \times 0.1 \times 1.562^2 = 731.95 \,\text{kVAr}$$

The total reactive power that is delivered by the grid is

$$Q_s = Q_1 + Q_m + Q_2 = 1.678 \, \text{MVAr}$$

The reactive power flow is opposite to the real power flow, as shown in Figure 8.10. In this case, the grid is the source for the reactive power.

The power factor *pf* at the terminals of the generator is

$$pf = \frac{P_s}{\sqrt{P_s^2 + Q_s^2}} = \frac{1.68}{\sqrt{1.68^2 + 1.678^2}} = 0.707 \, \text{leading}$$

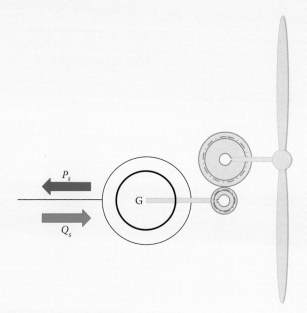

FIGURE 8.10
Real and reactive power flow of an induction generator.

Because of the high demand for reactive power by the squirrel-cage induction generator, wind farms with type 1 systems are required to generate their own reactive power locally. Some of these compensation systems are discussed in Chapter 12.

8.3.2 Inrush Current

At the moment when the generator is connected to the grid, its rotation speed could be different from the synchronous speed. Even for a small speed deviation, the current at the connection time could be excessive. This current is called "inrush currents," and its general shape is shown in Figure 8.11. The minimum inrush current occurs when the generator is switched into the grid while spinning at its synchronous speed. Any deviation from the synchronous speed could cause a substantial increase in the inrush current.

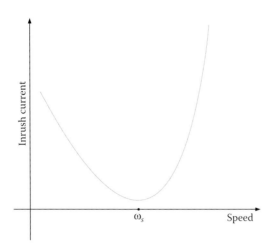

FIGURE 8.11
Inrush current of an induction machine.

EXAMPLE 8.8

For the system in Example 8.3, compute the rotor inrush current at starting. Assume the turbine is spinning the generator at 1350 r/min when the connecting switch is closed.

Solution:
Use Equation 8.13 to compute the rotor current at the time of connection. But first, let us compute the slip

$$s = \frac{n_s - n}{n_s} = \frac{1200 - 1350}{1200} = -0.125$$

$$I'_{a2} = \frac{V_{th}}{\sqrt{\left[r_{th} + \left(r'_2/s\right)\right]^2 + x_{eq}^2}} = \frac{379.39}{\sqrt{\left[0.0952 + \left(0.01/-0.125\right)\right]^2 + 0.109^2}} = 3.447\,\text{kA}$$

Keep in mind that $I'_{a2} = 1.562$ kA in normal operation. This means the increase in rotor current at starting is $(3.447/1.562) \times 100 = 221\%$. For the stator current, it is expected to increase by almost the same percentage.

To reduce the inrush current, the terminal voltage of the generator must be reduced at starting. This is done using an ac/ac converter between the stator and the grid. One of these converters is depicted in Figure 8.12. This system is called "soft starting." The main components of the converter are the anti-parallel silicon-controlled rectifiers (SCRs) and the bypass switch. Before the turbine is connected to the grid, the mechanical switch is in the open position. When the turbine spins the generator, the triggering angle of the SCRs is adjusted to reduce the voltage across the terminals of the generator. As seen in Chapter 5, the anti-parallel SCRs of each phase reduces the terminal voltage of the generator according to the relationship

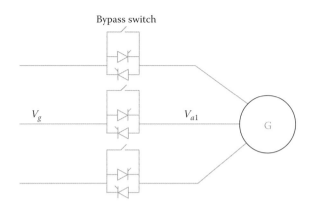

FIGURE 8.12
Soft starting of an induction generator.

$$V_{a1-st} = V_g \sqrt{\left(1 - \frac{\alpha}{\pi} + \frac{\sin 2\alpha}{2\pi}\right)} \qquad (8.32)$$

where:

V_g is the voltage at the grid side of the switch

V_{a1-st} is the reduced voltage of the stator at starting

α is the triggering angle of the forward SCR. The reverse SCR is triggered after 180° from α

After the generator is successfully connected to the grid, the triggering angle is reduced to slowly increase the terminal voltage of the generator. After it reaches its full value (rated V_{a1}), the mechanical switch is closed to bypass the SCRs.

EXAMPLE 8.9

For the system in Example 8.3, the speed of the generator is 1350 rpm when it is connected to the grid. Compute the terminal voltage of the generator that reduces the inrush current to 1 kA.

Solution:

Use Equation 8.13 to compute Thevinin's voltage.

$$I'_{a2-st} = \frac{V_{th-st}}{\sqrt{\left[r_{th} + \left(r'_2/s\right)\right]^2 + x^2_{eq}}} = \frac{V_{th-st}}{\sqrt{\left[0.0952 + \left(0.01/-0.125\right)\right]^2 + 0.109^2}} = 1.0\,\text{kA}$$

$$V_{th-st} = 110.05 \text{ V}.$$

Use Equation 8.9 to compute the reduced terminal voltage at starting V_{st}

$$V_{a1-st} = V_{th-st} \frac{\sqrt{r_1^2 + \left(x_1 + x_m\right)^2}}{x_m} = 110.05 \times \frac{\sqrt{0.01^2 + 2.1^2}}{2} = 115.55 \text{ V}$$

In line-to-line term

$$V_{a1-st-ll} = 115.55 \times \sqrt{3} = 200.15 \text{ V}$$

Voltage reduction

$$\Delta V_{a1-st} = \frac{200.15 - 690}{690} \times 100 = -71\%$$

EXAMPLE 8.10

For the case in Example 8.9, compute the triggering angle of the SCRs in Figure 8.12.

Solution:
Using Equation 8.32, the triggering angle α of ac/ac voltage control circuit in Figure 8.12 is

$$V_{a1-st} = V_g \sqrt{\left(1 - \frac{\alpha}{\pi} + \frac{\sin 2\alpha}{2\pi}\right)}$$

$$200.15 = 690 \sqrt{\left(1 - \frac{\alpha}{\pi} + \frac{\sin 2\alpha}{2\pi}\right)}$$

Solving for α iteratively yields

$$\alpha = 136.25°$$

8.3.3 Turbine Stability

The voltage of the power grid is suppressed during faults, and the largest reduction occurs near the area of the fault. The depression in voltage reduces the electrical power generated by the turbine for as long as the fault lasts. From the turbine viewpoint, the reduction of its electrical power output must be matched by a quick reduction in the mechanical power captured by the blades. Otherwise, the turbine will accelerate and the speed of the blades could reach a destructive level. During the period from the initiation of the fault and until the blades are pitched to the correct angle, the mechanical power going into the generator P_m is higher than the electrical power output of the generator P_{out}. If we ignore the losses of the windings, the power imbalance of the generator is

$$P_m > P_{out} \tag{8.33}$$

From the energy viewpoint, we can write Equation 8.33 as

$$\int_0^t P_m \, dt > \int_0^t P_{out} \, dt \tag{8.34}$$

$$E_m > E_{out}$$

where:
  t is the time during which the fault is present
  E_m is the energy provided by the turbine blades during fault
  E_{out} is the output energy of the generator during fault

Because the input energy to the generator is larger than the output energy, the difference is stored in the rotating mass in the form of kinetic energy, which is expressed by

$$\Delta KE = E_m - E_{\text{out}} = \frac{1}{2} J \frac{d\omega_2}{dt} \tag{8.35}$$

where:
   KE is the kinetic energy of rotating mass (rotor of generator, gearbox, shaft, hub, and blades)
   J is the inertia constant of the rotating mass seen from the generator side of the gearbox
   ω_2 is the speed of the generator

During the steady state, $\Delta KE = 0$ and the generator is spinning at constant speed. During fault, $\Delta KE > 0$, causing the speed of the generator to increase due to the acceleration in Equation 8.35. This could result in the blades speeding up to a level that could damage the turbine.

To restore the balance of energy, we have essentially three options, which are as follows:

1. Option 1—Reduce the mechanical power quickly. However, the mechanical power cannot be changed fast enough for the cases of severe faults. Pitching the blades takes some time, as it involves rapid motion of large masses (blades, gearbox, generator, etc.). In addition, rapid actuation may cause the blades to wobble and could hit the tower.

2. Option 2—Increase the electrical power output. This option can be implemented by two methods:

   a. Store the extra energy ΔKE somewhere outside the rotating mass.

   b. Dissipate the extra energy in external circuits.

3. Option 3—A combination of option 1 and option 2.

Storing energy during disturbances is an expensive option. It requires the use of devices that can acquire burst of energy in short time such as super capacitor and flywheel. Dissipating the extra energy during transients is a more economical option. One of the effective techniques is the stator dynamic resistance (SDR).

The SDR system, shown in Figure 8.13, consists of three-phase resistance bank (R_{SDR}) connected between the generator terminals and the grid. The resistance is shunted by a bidirectional solid-state switch and a bypass mechanical switch. When the system is operating normally, the bypass switch is closed and no energy is consumed by the resistance, R_{SDR}. When a fault occurs, the bypass switch is opened and the solid-state switch regulates the amount of energy absorbed by the resistance. Because the circuit breakers in the power system clears faults within a few cycles, the SDR operates for just a few milliseconds.

The energy absorbed by the resistance bank is the amount needed to maintain the power balance until the mechanical power is appropriately reduced. The instantaneous power consumed by the resistance bank is

$$p_{\text{SDR}} = 3 \, i^2 \, R_{\text{SDR}} \tag{8.36}$$

where:

$$i = I_{\text{max}} \sin \omega t \tag{8.37}$$

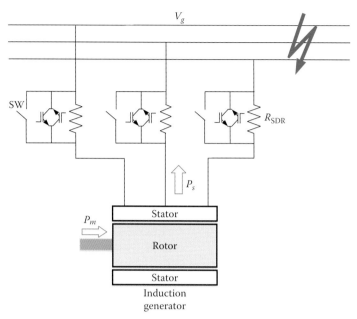

FIGURE 8.13
Stator dynamic resistance.

The actual power consumed by the resistance is the average of the instantaneous power during the period when the solid-state switches are open.

$$P_{SDR} = \frac{3}{\pi} \left[\int_0^\alpha i^2 R_{SDR} \, d\omega t + \int_\beta^\pi i^2 R_{SDR} \, d\omega t \right]$$

(8.38)

$$P_{SDR} = \frac{3 R_{SDR}}{\pi} \left[\int_0^\alpha I_{max}^2 \sin^2 \omega t \, d\omega t + \int_\beta^\pi I_{max}^2 \sin^2 \omega t \, d\omega t \right]$$

$$P_{SDR} = \frac{3 I_{a1}^2 R_{SDR}}{\pi} \left(\pi - \gamma - \frac{\sin 2\alpha}{2} + \frac{\sin 2\beta}{2} \right)$$

(8.39)

where:
I_{a1} is the rms current of the stator
α is the angle at which the solid-state switch is closed in one half of the ac cycle
β is the angle at which the solid-state switch is opened in one half of the ac cycle
γ is the conduction period of the solid-state switch

$$\gamma = \beta - \alpha$$

(8.40)

In Equation 8.39, define the effective resistance R_e as

$$R_e = \frac{R_{SDR}}{\pi} \left(\pi + \gamma - \frac{\sin 2\alpha}{2} + \frac{\sin 2\beta}{2} \right)$$

(8.41)

Hence, the power consumed by the resistance is

$$P_{SDR} = 3 I_{a1}^2 R_e \qquad (8.42)$$

To balance the power, P_{SDR} should compensate for the drop in the output power of the generator ΔP_{out}

$$P_{SDR} = -\Delta P_{out} \qquad (8.43)$$

To compute the energy level of R_{SDR}, we need to know the average current of the generator during fault. Because the current is widely changing during fault, we can approximate the calculations by assuming the current in Equation 8.42 to be the rms value of the steady-state fault current, I_{f-ss}.

$$P_{SDR} = 3 I_{f-ss}^2 R_e \qquad (8.44)$$

EXAMPLE 8.11

A fault occurs near type 1 wind turbine causing the voltage at the grid side to substantially reduce. The fault lasts for 100 ms. During fault, the average deficit in energy is 300 kJ. Compute the value of the effective resistance to be inserted between the generator and the grid during fault that limits the current of the generator to 1.0 KA.

Solution:
The power to be consumed by the effective resistance is

$$P_{SDR} = \frac{E_{SDR}}{t} = \frac{300}{0.1} = 3.0\,\text{MW}$$

R_e can be computed using Equation 8.44

$$R_e = \frac{P_{SDR}}{3 I_{f-ss}^2} = \frac{3 \times 10^6}{3 \times 10^6} = 1.0\,\Omega$$

EXAMPLE 8.12

If the SDR resistance is 5Ω, compute the triggering and commutation angles for the case in Example 8.11

Solution:
Using Equation 8.41 yields

$$\frac{R_e}{R_{SDR}} = \frac{1}{5} = \frac{1}{\pi}\left(\pi - \gamma - \frac{\sin 2\alpha}{2} + \frac{\sin 2\beta}{2}\right)$$

Select the closing angle of the solid-state switches, say $\alpha = 30°$. Then we can solve for the opening angle β

$$-2.603 = -\beta + \frac{\sin 2\beta}{2}$$

By iteration, $\beta = 123°$

Exercise

1. What is the difference between an airgap and developed torques?
2. Is type 1 turbine a constant or a variable-speed system?
3. What is pullout power?
4. What are the factors that determine the maximum power of a type 1 generator?
5. What are the factors that determine the maximum torque of a type 1 generator?
6. Is the speed at maximum torque dependent on terminal voltage?
7. What are the advantages and disadvantages of type 1 turbines?
8. Can type 1 generator produce its own reactive power?
9. What causes inrush current?
10. How is inrush current reduced?
11. What causes the generator to become unstable?
12. How is the generator controlled during grid faults?
13. A type 1 wind turbine has a six-pole, 60 Hz three-phase, Y-connected induction generator. The speed of the generator is 1250 r/min. The terminal voltage of the generator is 690 V. The parameters of the machine are $r_1 = r_2' = 10 \text{ m}\Omega$; $x_1 = x_2' = 100 \text{ m}\Omega$; and $x_m = 2 \ \Omega$.

    Ignore the core losses and the rotational losses of the generator. Compute the following:

    a. Developed power
    b. Developed torque
    c. Speed at maximum torque
    d. Maximum torque
    e. Speed at maximum power
    f. Maximum power

14. A type 1 wind turbine has a six-pole, 60 Hz three-phase, Y-connected induction generator. The speed of the generator is 1260 r/min. The terminal voltage of the generator is 415 V. The parameters of the machine are: $r_1 = r_2' = 4.0 \text{ m}\Omega$; $x_1 = x_2' = 40 \text{ m}\Omega$; and $x_m = 5 \ \Omega$.

    Compute the reactive power delivered by the grid to the generator.

15. A type 1 wind turbine has a six-pole, 60 Hz three-phase, Y-connected induction generator. The terminal voltage of the generator is 415 V. The parameters of the machine are $r_1 = r_2' = 4.0\,m\Omega$; $x_1 = x_2' = 40\,m\Omega$; and $x_m = 5\,\Omega$.

    Assume the turbine is spinning the generator at 1260 r/min when the connecting switch is closed. Compute the rotor inrush current at starting. Also, compute the triggering angle to achieve soft starting, whereby the initial inrush current is limited to 20% of the rated current.

16. Compute the triggering angles of a stator dynamic resistance bank that consumes 900 kJ in 50 ms. Assume the SDR resistance is 50 Ω and the steady-state fault current of the generator is 500 A.

17. A fault near type 1 wind turbine lasts for 100 ms. Compute the triggering angles of the 2 Ω SDR resistance that keeps the steady-state generator current to 1.0 kA and consumes 300 kJ.

9

Type 2 Wind Turbine System

The induction generator used in type 2 turbines is mostly the wound rotor (slip ring) generator. The rotor of this generator is connected to adjustable resistance through power electronic switching devices, one of these systems is shown in Figure 9.1. Each winding is connected to an external resistance r, with two solid-state switches in anti-parallel configuration in shunt with the resistance. When the switches are continuously closed, the resistances are shorted. When the full value of the resistances is needed, the switches are continuously open. If the value of the resistances needs to be adjusted, the duty ratio of the switches is regulated. To obtain a smooth operation, the switching frequency of the solid-state devices is set high (kHz range).

The single-phase representation of the rotor circuit is shown in Figure 9.2. Because the switching frequency of the transistors is very high, we can assume that the rotor voltage (which is very low frequency) is unchanged during the single-switching period τ, in Figure 9.3. t_{on} in the figure is the closing interval of the transistor and t_{off} is the opening interval. During the closing interval, the external resistance seen by the rotor is zero and during the opening interval, the resistance is r.

The added resistance to the rotor circuit (r_{add}) is the average during one switching period.

$$r_{add} = \frac{1}{\tau} \int_{t_{on}}^{\tau} r \, dt = r \frac{\tau - t_{on}}{\tau} = r(1 - k) \tag{9.1}$$

where:
k is the duty ratio of the transistor

$$k = \frac{t_{on}}{\tau} \tag{9.2}$$

9.1 Equivalent Circuit of Type 2 Generator

To understand the impact of the external resistance on the operation of the induction generator, examine the induced voltage in the rotor windings. At standstill, the induced voltage of the rotor is E_{a2}, which is proportional to the speed of the airgap field n_s.

$$E_{a2} \sim n_s \tag{9.3}$$

When the rotor is spinning, the induced voltage in the rotor windings E_r is proportional to the difference in speed between the rotating magnetic field in the airgap and the speed of the rotor.

$$E_r \sim \Delta n = n_s - n \tag{9.4}$$

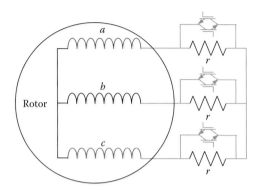

FIGURE 9.1
Adding resistance to the rotor circuit through a high-frequency converter.

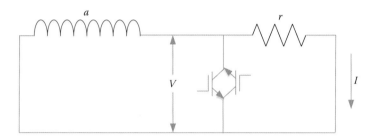

FIGURE 9.2
Single-phase rotor circuit with external resistance.

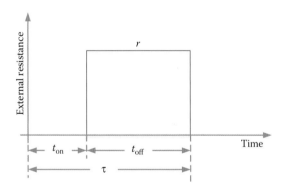

FIGURE 9.3
Effective resistance.

Hence,

$$\frac{E_r}{E_{a2}} = \frac{n_s - n}{n_s} = s \tag{9.5}$$

In other words, the induced voltage across the rotor windings is proportional to the slip

$$E_r = s E_{a2} \tag{9.6}$$

Similarly, the frequency of the rotor current is directly proportional to the rate by which the magnetic field in the airgap cuts the rotor winding. At standstill, the airgap flux cuts the rotor windings at a rate proportional to the synchronous speed n_s. Hence, the frequency of the rotor current at standstill f_s and is proportional to the synchronous speed. This makes f_s equal to the frequency of the grid f.

$$f = f_s \sim n_s \tag{9.7}$$

If the rotor is spinning at a speed n, the flux cuts the rotor windings at a rate proportional to Δn. Hence, the frequency of the rotor current f_r at any speed is proportional to Δn.

$$f_r \sim \Delta n = n_s - n \tag{9.8}$$

Hence,

$$\frac{f_r}{f} = \frac{n_s - n}{n_s} = s \tag{9.9}$$

$$f_r = sf \tag{9.10}$$

Equation 9.10 shows that the frequency of the rotor current is proportional to the slip. This makes the rotor frequency changing with speed.

The inductive reactance of the rotor windings x_r at any rotor frequency is

$$x_r = 2\pi f_r L_2 \tag{9.11}$$

Substituting the value of f_r in Equation 9.10 into Equation 9.11 yields

$$x_r = 2\pi(sf)L_2 = sx_2 \tag{9.12}$$

where:
x_2 is the rotor inductive reactance at standstill (i.e., at grid frequency)

$$x_2 = 2\pi f L_2 = \omega L_2 \tag{9.13}$$

Based on Equations 9.6 and 9.12, we can obtain the single-phase equivalent circuit of the rotor shown in Figure 9.4a. The rotor current I_{a2} in the circuit is

$$\overline{I}_{a2} = \frac{s\overline{E}_{a2}}{(r_2 + r_{\text{add}}) + jsx_2} = \frac{\overline{E}_{a2}}{[(r_2 + r_{\text{add}})/s] + jx_2} \tag{9.14}$$

To separate the losses in the rotor circuit from the developed power, the resistive component in Equation 9.14 can be divided into two parts.

$$\frac{r_2 + r_{\text{add}}}{s} = (r_2 + r_{\text{add}}) + \frac{r_2 + r_{\text{add}}}{s}(1 - s) = r_{\text{eq2}} + r_{d-\text{add}} \tag{9.15}$$

where:
$r_{\text{eq2}} = (r_2 + r_{\text{add}})$ is the rotor winding resistance plus the added resistance
$r_{d-\text{add}} = \left[(r_2 + r_{\text{add}})/s\right](1 - s)$ is the developed power resistance

The equivalent circuit is then modified, as shown in Figure 9.4b. After adding Thevinin's circuit for the stator and referring all parameters to the stator circuit using the effective turns ratio, we can obtain the equivalent circuit in Figure 9.5, where

$$\overline{V}_{D-\text{add}} = -r'_{d-\text{add}} \overline{I}'_{a2} = -\frac{r'_2 + r'_{\text{add}}}{s}(1 - s)\overline{I}'_{a2} \tag{9.16}$$

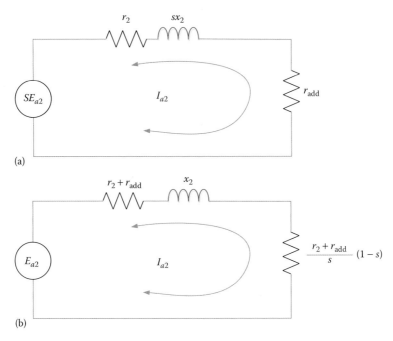

FIGURE 9.4
(a) Rotor equivalent circuit without added resistance and (b) rotor equivalent circuit with added resistance.

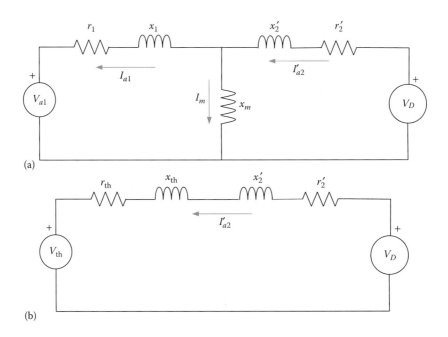

FIGURE 9.5
(a) Generator equivalent circuit without added resistance and (b) generator equivalent circuit with added resistance.

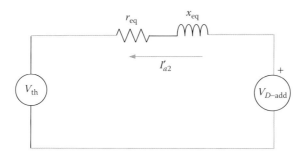

FIGURE 9.6
Modified equivalent circuit for induction generator with added resistance to rotor circuit.

The equivalent circuit can further be modified by aggregating the parameters, as shown in Figure 9.6, where

$$x_{eq} = x_{th} + x_2' \tag{9.17}$$

$$r_{eq} = r_{th} + r_2' + r_{add}' \tag{9.18}$$

The rotor current of the generator can be computed using the modified equivalent circuit in Figure 9.6.

$$\overline{I}_{a2}' = \frac{\overline{V}_{D-add} - \overline{V}_{th}}{r_{eq} + jx_{eq}} = \frac{-\overline{V}_{th}}{\{r_{th} + [(r_2' + r_{add}')/s]\} + jx_{eq}} \tag{9.19}$$

9.2 Real Power

The developed power of the generator is

$$P_d = 3V_D I_{a2}' = -3\frac{r_2' + r_{add}'}{s}(1-s)(I_{a2}')^2 \tag{9.20}$$

Substituting the value of the current in Equation 9.19 into Equation 9.20 yields

$$P_d = -3\frac{(1-s)}{s}\frac{V_{th}^2(r_2' + r_{add}')}{\{r_{th} + [(r_2' + r_{add}')/s]\}^2 + x_{eq}^2} \tag{9.21}$$

The pullout (maximum) power and the slip at the pullout power can be computed by equating the derivative of 9.21 to zero with respect to the slip.

$$\frac{dP_d}{ds} = -\frac{d}{ds}\left[3\frac{(1-s)}{s}\frac{V_{th}^2(r_2' + r_{add}')}{\{r_{th} + [(r_2' + r_{add}')/s]\}^2 + x_{eq}^2}\right] = 0 \tag{9.22}$$

The solution leads to the slip at pullout power (s'), which is

$$s' = -m \left[m + \sqrt{m^2 + 1} \right] \tag{9.23}$$

where

$$m = \frac{r_2' + r_{add}'}{\sqrt{r_{th}^2 + 2r_{th}(r_2' + r_{add}') + x_{eq}^2}} \tag{9.24}$$

Substituting Equation 9.23 into Equation 9.21 yields the equation for the pullout power (maximum power)

$$P_{d-max} = -3 \frac{(1-s')}{s'} \frac{V_{th}^2 (r_2' + r_{add}')}{\{r_{th} + [(r_2' + r_{add}')/s']\}^2 + x_{eq}^2} \tag{9.25}$$

Note that the pullout power is a function of the added resistance in the rotor circuit.

Based on Equation 9.21, the power–speed characteristics of type 2 turbine can be plotted, as shown in Figure 9.7. When the duty ratio of the switching circuit decreases, the value of r_{add}' increases. This causes the following:

- Slope of the characteristic decreases.
- Magnitude of the pullout power increases.
- Operating range of speed increases from $\Delta\omega_1$ to $\Delta\omega_2$. It is possible for $\Delta\omega_2$ to have a value of up to 10% of the synchronous speed.
- Range of developed power increases from P_1 to P_2.

In addition, because the added resistance can be adjusted quite rapidly, type 2 system can regulate its production. Take, for instance, the cases in Figures 9.8 and 9.9. In Figure 9.8, we assume that the output power of the generator need to be maintained constant even when the speed changes. Ignore the losses in the system and assume that the initial operating

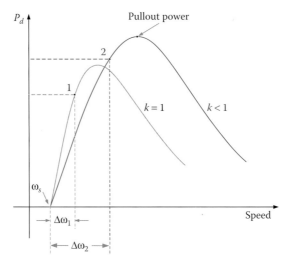

FIGURE 9.7
Power–speed characteristics of a type 2 system.

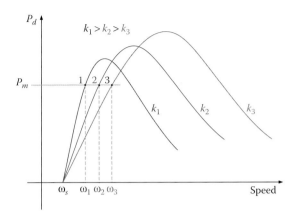

FIGURE 9.8
Power regulation of a type 2 system: constant output power.

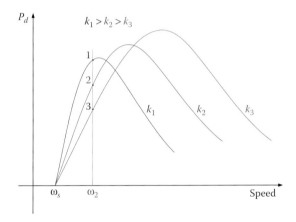

FIGURE 9.9
Power regulation of a type 2 system: adjustable output power.

point is point 2, where the speed is ω_2. If the wind speed is reduced to ω_1 or increases to ω_3, the output power can be maintained constant by adjusting the duty ratio of the switching devices to operate at either point 1 or 3, respectively.

In Figure 9.9, we assume that the output power needs to be regulated even when the speed is unchanged. In this case, point 2 (initial operating point) can increase to point 1 or decreased to point 3 by adjusting the duty ratio accordingly.

EXAMPLE 9.1

A type 2 wind turbine has a six-pole, 60 Hz three-phase, Y-connected induction generator. The terminal voltage of the generator is 690 V. The parameters of the machine are $r_2' = 10\,\text{m}\Omega$; $x_2' = 100\,\text{m}\Omega$; and $x_m = 5\,\Omega$.

Assume the impedance of the stator circuit is much smaller than the impedance of the magnetizing branch. Compute the pullout power. Repeat the solution after inserting a $0.02\,\Omega$ resistance in the rotor circuit as referred to the stator. The duty ratio of its solid-state switching circuit is 50%.

Solution:

If we ignore the stator parameters, the following approximation can be made:

$$V_{th} \approx V_{a1}$$

$$r_{th} \approx 0$$

$$x_{eq} \approx x_2'$$

Without the added rotor resistance, we compute the slip at pullout power as given in Equations 9.23 and 9.24

$$m = \frac{r_2'}{x_2'} = 0.1$$

$$s' = -m\left[m + \sqrt{m^2 + 1}\right] = -0.1 \times \left[0.1 + \sqrt{0.1^2 + 1}\right] = -0.1105$$

The speed at the pullout power is

$$n' = n_s(1 - s') = 1200 \times (1 + 0.1105) = 1332.6 \text{ rpm}$$

The speed range to pullout power is

$$\Delta n' = n_s - n' = 1200 - 1332.6 = -132.6 \text{ rpm}$$

The pullout power (maximum power) is

$$P_{d-\max} = -3\frac{(1-s')}{s'}\frac{V_{a1}^2 r_2'}{(r_2'/s')^2 + (x_2')^2} = 3\frac{(1+0.1105)}{0.1105}\frac{(690^2/3)\times 0.01}{(0.01/-0.1105)^2 + (0.1)^2} = 2.53 \text{ MW}$$

When the resistance is added, we need to compute r_{add} based on the duty ratio in Equation 9.1

$$r_{add} = r(1 - k) = 0.02 \times (1 - 0.5) = 0.01\Omega$$

Now we can compute the slip at pullout power

$$m = \frac{r_2' + r_{add}'}{x_2'} = \frac{0.01 + 0.01}{0.1} = 0.2$$

$$s' = -m\left[m + \sqrt{m^2 + 1}\right] = -0.2 \times \left[0.2 + \sqrt{0.2^2 + 1}\right] = -0.244$$

The speed at the pullout power is

$$n' = n_s(1 - s') = 1200 \times (1 + 0.244) = 1492.8 \text{ rpm}$$

The speed range to pullout power is

$$\Delta n' = n_s - n' = 1200 - 1492.8 = 292.8 \text{ rpm}$$

The pullout power is

$$P_{d-\max} = -3\frac{(1-s')}{s'}\frac{V_{a1}^2(r_2' + r_{add}')}{\left\lceil(r_2' + r_{add}')/s'\right\rceil^2 + (x_2')^2} = 3\frac{(1+0.244)}{0.244}\frac{(690^2/3)0.02}{(0.02/-0.244)^2 + (0.1)^2} = 2.904 \text{ MW}$$

Note that the speed range of the generator with added resistance to the rotor is wider than without it. Also, the pullout power increases with the added resistance.

EXAMPLE 9.2

For the system in Example 9.1, assume the generator is operating at a speed of 1230 rpm. If wind speed increases the generator speed to 1260 rpm, compute the added resistance that maintains the developed power constant. Also, compute the power consumed by the added resistance.

Solution:
Without the added rotor resistance, the slip is

$$s = \frac{n_s - n}{n_s} = \frac{1200 - 1230}{1200} = -0.025$$

The developed power can be obtained from Equation 9.21 by setting $r'_{add} = 0$.

$$P_d = -3\frac{(1-s)}{s}\frac{V_{a1}^2(r'_2)}{(r'_2/s)^2 + (x'_2)^2} = 1.148\,\text{MW}$$

The slip at the new speed can be computed as

$$s = \frac{n_s - n}{n_s} = \frac{1200 - 1260}{1200} = -0.05$$

To compute the added resistance, we can use Equation 9.21

$$P_d = -3\frac{(1-s)}{s}\frac{V_{a1}^2(r'_2 + r'_{add})}{\left[(r'_2 + r'_{add})/s\right]^2 + (x'_2)^2}$$

$$1.148 \times 10^6 = -\frac{(1+0.05)}{-0.05}\frac{690^2(0.01 + r'_{add})}{\left[(0.01 + r'_{add})/-0.05\right]^2 + (0.1)^2}$$

The solution of the above equation gives r'_{add}

$$r'_{add} = 0.0218\,\Omega$$

To compute the power dissipated in the added resistance, we need to compute the rotor current using Equation 9.19.

$$\bar{I}'_{a2} \approx \frac{-\bar{V}_{a1}}{\left[(r'_2 + r'_{add})/s\right] + jx'_2} = \frac{-\left(690/\sqrt{3}\right)}{(0.0318/-0.05) + j0.1} = 618.77\angle -8.93°\,\text{A}$$

The dissipated power in the added resistance bank is

$$P_{add} = 3\,r'_{add}\left(I'_{a2}\right)^2 = 25.04\,\text{kW}$$

9.3 Electric Torque

The maximum electric (developed) torque can be obtained by similar process to the one used for the maximum power. The developed torque is

$$T_d = \frac{P_d}{\omega_2} = \frac{P_d}{\omega_s(1-s)} = -\frac{3}{s\omega_s}\frac{V_{th}^2(r'_2 + r'_{add})}{\left\{r_{th} + [(r'_2 + r'_{add})/s]\right\}^2 + x_{eq}^2} \tag{9.26}$$

The maximum torque is obtained by equating the derivative of the torque equation to zero with respect to the slip.

$$\frac{dT_d}{ds} = -\frac{d}{ds}\left[\frac{3}{s\omega_s}\frac{V_{th}^2(r_2' + r_{add}')}{\left\{r_{th} + [(r_2' + r_{add}')/s]\right\}^2 + x_{eq}^2}\right] = 0 \tag{9.27}$$

The solution of Equation 9.27 leads to the slip at maximum torque s^*

$$s^* = -\frac{r_2' + r_{add}'}{\sqrt{r_{th}^2 + x_{eq}^2}} \tag{9.28}$$

The negative sign in Equation 9.28 is for generators at super-synchronous speed. Substituting the value of s^* in Equation 9.28 into Equation 9.26 yields

$$T_{d\text{-max}} = -\frac{3}{s^*\omega_s}\frac{V_{th}^2(r_2' + r_{add}')}{\left\{r_{th} + [(r_2' + r_{add}')/s^*]\right\}^2 + x_{eq}^2}$$

$$T_{d\text{-max}} = \frac{3V_{th}^2}{2\omega_s\left(\sqrt{r_{th}^2 + x_{eq}^2} - r_{th}\right)} \tag{9.29}$$

Note that the magnitude of the maximum torque ($T_{d\text{-max}}$) is not dependent on the value of the added resistance. Based on Equation 9.26, the torque–speed characteristics can be

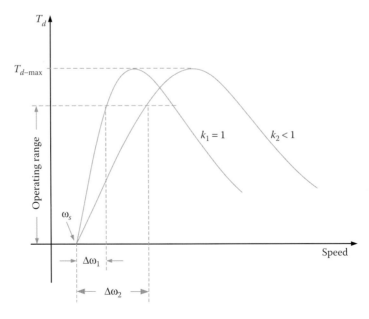

FIGURE 9.10
Developed torque–speed characteristics of a type 2 system.

plotted, as shown in Figure 9.10. When the duty ratio changes, the speed at which the maximum torque occurs changes as well. For the operating range of the electrical torque, the insertion of the resistance in the rotor circuit increases the range of the operating speed $\Delta\omega$.

EXAMPLE 9.3

For the system in Example 9.1, compute the maximum torque and the speed at maximum torque without and with the added resistance.

Solution:
Without the added rotor resistance, we compute the slip at maximum torque using Equation 9.28.

$$s^* = -\frac{r_2'}{x_2'} = \frac{0.01}{0.1} = -0.1$$

The speed at the maximum torque is

$$n^* = n_s(1-s^*) = 1200 \times (1+0.1) = 1320\,\text{rpm}$$

The speed range to maximum torque is

$$\Delta n^* = n_s - n^* = 1200 - 1320 = 120\,\text{rpm}$$

The maximum torque is

$$T_{d-\text{max}} = \frac{3}{2}\frac{V_{a1}^2}{x_2'\omega_s} = \frac{690^2}{2 \times 0.1 \times 40\pi} = 18.94\,\text{kNm}$$

With the added resistance, the slip at maximum torque is

$$s^* = -\frac{r_2' + r_{\text{add}}'}{x_2'} = \frac{0.01+0.01}{0.1} = -0.2$$

The speed at the maximum torque is

$$n^* = n_s(1-s') = 1200 \times (1+0.2) = 1440\,\text{rpm}$$

The speed range to maximum torque is

$$\Delta n^* = n_s - n^* = 1200 - 1440 = 240\,\text{rpm}$$

The maximum torque is the same as it is not a function of the added resistance

$$T_{d-\text{max}} = \frac{3}{2}\frac{V_{a1}^2}{x_2'\omega_s} = \frac{690^2}{2 \times 0.1 \times 40\pi} = 18.94\,\text{kNm}$$

Note that the speed at pullout power n' is not the same as the speed at maximum torque n^*

9.4 Assessment of Type 2 System

Type 2 wind turbine system is more flexible than type 1: the operating ranges for power and speed are higher than those for type 1. Type 2 generators normally require access to the rotor through a slip-ring system, which makes the machine more expensive and higher maintenance. However, newer systems have the resistances installed in the rotating mass of the generator, whereby its converter is triggered using optical signals from the stationary stator.

Similar to type 1 system, the main drawback of type 2 turbines is their lack of reactive power control and voltage support. In wind farms with types 1 or 2 turbines, the voltage support must be implemented by external equipment that is capable of adaptive reactive power production. These systems are discussed in Chapter 12. Another drawback of type 2 system is its low efficiency because of the power consumption in the added resistance.

9.5 Control and Protection of Type 2 System

The control and protection of type 2 system is similar to that for type 1. The reactive power problem and its solution for type 2 turbine is the same as that for type 1, which is discussed in Chapter 8. The other two problems, namely, inrush current and turbine stability, can be addressed by the added resistance in the rotor circuit.

9.5.1 Inrush Current

The inrush current occurs at the initial connection of the turbine to the grid. The starting sequence is initiated by the release of the brake to allow the turbine to spin freely. When the speed is near the synchronous speed, the turbine is connected to the grid. Because the speed of the turbine at the initial connection to the grid cannot preciously equal the synchronous speed, an inrush current flows through the generator. In type 1 turbine, the stator dynamic resistance is used to limit the inrush current. In type 2, the inrush current can be limited by the added rotor resistance.

The general expression for the rotor current is given in Equation 9.19. The inrush current in the rotor circuit can be computed by setting the slip to the value at the time of grid connection. If we make an approximation, whereby the winding impedance of the stator is ignored, the rotor starting current is then

$$\overline{I}'_{a2-st} = \frac{-\overline{V}_{a2}}{\left[(r'_2 + r'_{add})/s\right] + jx'_2} \tag{9.30}$$

Using Figure 9.5, we can compute the stator current at starting

$$\overline{I}'_{a1-st} = \overline{I}'_{a2-st} - \overline{I}_m = \frac{-\overline{V}_{a2}}{\left[(r'_2 + r'_{add})/s\right] + jx'_2} - \frac{\overline{V}_{a2}}{jx_m} \tag{9.31}$$

EXAMPLE 9.4

For the system in Example 9.1, the turbine is connected to the grid while spinning at 1230 rpm. Compute the inrush current in the stator windings. Repeat the solution when 0.03Ω resistance (refered to stator) is inserted in the rotor circuit.

Solution:

The slip at the grid interconnection is

$$s = \frac{1200 - 1230}{1200} = -0.025$$

Using Equation 9.30, we can compute the inrush current in the rotor without the added resistance

$$\bar{I}'_{a2-st} = \frac{-\bar{V}_{a2}}{\left(r'_2/s\right) + jx'_2} = \frac{-\left(690/\sqrt{3}\right)}{\left(0.01/-0.025\right) + j0.1} = 966.2 \angle 14.04° \text{A}$$

The current in the magnetizing branch is

$$\bar{I}_m = \frac{\bar{V}_{a2}}{jx_m} = \frac{690/\sqrt{3}}{j5} = j79.67 \text{ A}$$

The stator inrush current is

$$\bar{I}_{a1-st} = \bar{I}'_{a2-st} - \bar{I}_m = 966.2 \angle 14.04° - j79.67 = 950.0 \angle 9.37° \text{ A}$$

With the added resistance, the rotor inrush current is

$$\bar{I}'_{a2-st} = \frac{-\bar{V}_{a2}}{\left[\left(r'_2 + r'_{add}\right)/s\right] + jx'_2} = \frac{-\left(690/\sqrt{3}\right)}{\left[\left(0.01 + 0.03\right)/-0.025\right] + j0.1} = 248.5 \angle 3.58° \text{A}$$

The stator inrush current with the added resistance is

$$\bar{I}_{a1-st} = \bar{I}'_{a2-st} - \bar{I}_m = 248.5 \angle 3.58° - j79.67 = 265.64 \angle 21° \text{A}$$

Note that the inrush current is reduced substantially when the rotor resistance is added.

9.5.2 Turbine Stability

As explained in Chapter 8, during voltage depression, the output power of the turbine is reduced. For stable operation, the mechanical power from the turbine must be reduced to match the output electrical power plus all losses. However, the reduction of the mechanical power takes time as it involved the control of large mass. During this period, the extra energy acquired by the rotating mass must be dissipated to prevent the machine from damaging overspending.

The added resistance in the rotor circuit can help consume part of this extra energy. Examine the next example.

EXAMPLE 9.5

For the system in Example 9.1, assume the turbine is producing 2 MW at a steady-state speed of 1230 rpm. A nearby fault reduces the terminal voltage of the generator to 80% of its steady-state value. Compute the power consumed by 0.03Ω added resistance (refereed to stator) and the rotor current referred to stator. Also, compute the energy consumed by the resistance if the voltage depression lasts for 50 ms.

Solution:
The slip at the initiation of the fault is

$$s = \frac{1200 - 1230}{1200} = -0.025$$

If we ignore the electrical losses of the generator, the output power is the same as the developed power. Equation 9.21, shows that the developed power is proportional to the square of the voltage. If we ignore the parameters of the stator windings, the developed power without the added rotor resistance is

$$P_{d1} = -3\frac{(1-s)}{s}\frac{V_{a1}^2 r_2'}{(r_2'/s)^2 + x_2'^2} = 2\,\text{MW}$$

The power with voltage depression and added rotor resistance is

$$P_{d2} = -3\frac{(1-s)}{s}\frac{(0.8 \times V_{a1})^2(r_2' + r_{add}')}{\left[(r_2' + r_{add}')/s\right]^2 + x_2'^2}$$

Hence,

$$P_{d2} = P_{d1}\frac{(0.8)^2(r_2' + r_{add}')\left[(r_2'/s)^2 + x_2'^2\right]}{r_2'\left\{[(r_2' + r_{add}')/s] + x_2'^2\right\}} = 2\frac{(0.8)^2(0.01 + 0.03)\left[(0.01/-0.025)^2 + 0.01^2\right]}{0.01 \times \left\{[0.01 + 0.03/-0.025]^2 + 0.01^2\right\}}$$

$$P_{d2} = 320.18\,\text{kW}$$

The power that needs to be consumed by the added resistance is

$$P_{add} = P_{d1} - P_{d2} = 2.0 - 0.32018 = 1.679\,\text{MW}$$

The rotor current is

$$I_{a2}' = \sqrt{\frac{P_{add}}{r_{add}'}} = \sqrt{\frac{(1.679/3)}{0.03}} = 4.32\,\text{kA}$$

The maximum energy consumed by the three-phase added resistance is

$$E_{add} = P_{add}t = 1.679 \times 0.05 = 83.95\,\text{kJ}$$

Exercise

1. What are the main differences between type 1 and type 2 turbines?

2. What are the advantages and disadvantages of type 2 turbines?

3. Is type 2 turbine constant or variable-speed system?

4. Does the effective rotor resistance increases or decreases with the increase of the duty ratio?

5. What are the factors that determine the maximum power of a type 2 generator?

6. What are the factors that determine the maximum torque of a type 2 generator?

7. Is the speed at maximum torque dependent on the terminal voltage?

8. Can a type 2 generator produce its own reactive power?

9. What causes inrush current?

10. How is inrush current reduced?

11. What causes the generator to become unstable?

12. How is the generator controlled during grid faults?

13. How is the pullout torque adjusted?

14. A type 2 wind turbine has blades of 50 m long. The induction generator of the turbine is a six-pole machine. The inductive reactance of the core is 3.0Ω. The rotor resistance and inductive reactance referred to the stator are 0.01Ω and 0.1Ω, respectively. The stator windings impedances are much smaller than the core inductive reactance. Because of an increase in wind speed, the generator speed increases from 1260 to 1290 rpm. Compute the added resistance to the rotor circuit to maintain the lift force constant.

15. A type 2 wind turbine has six-pole, 60 Hz three-phase, Y-connected induction generator. The terminal voltage of the generator is 690 V. The rotor parameters of the machine are $r_2' = 10\,\text{m}\Omega$; $x_2' = 100\,\text{m}\Omega$; and $N_1/N_2 = 2$.

   a. Ignore the impedance of the stator winding and compute the speed at maximum torque.

   b. Assume a solid-state switching circuit is used to regulate 0.05Ω resistance in the rotor circuit. Compute the duty ratio of the switching circuit that makes the maximum torque occur at 10% higher speed.

16. A 690 V generator of type 2 turbine has the following parameters: $r_2' = 10\,\text{m}\Omega$; $x_2' = 100\,\text{m}\Omega$; and $N_1/N_2 = 2$.

   The turbine was producing 1.0 MW at a steady-state speed of 1230 rpm when the grid voltage was depressed by 20%. Ignore the electrical losses and compute the value of the added resistance that would limit the rotor current to 2.0 kA.

17. A 1.5 MVA, type 2 wind turbine has eight-pole, 60 Hz three-phase, Y-connected induction generator. The terminal voltage of the generator is 690 V. When the machine was running at 936 rpm, the output power was 1.0 MW. After a voltage depression on the grid, the output power of the generator is reduced by 30%. Compute the value of the added resistance that would limit the rotor current to 150% of the rated current of the stator. You may ignore the losses of the generator.

18. A type 2 wind turbine has 1Ω added resistance installed in its rotor circuit. The resistance is switched by anti-parallel switching circuit in shunt with the resistance. At a given disturbance, the value of the added resistance needs to be 0.5Ω. Compute the triggering angle of the switching circuit.

10

Type 3 Wind Turbine System

To date, type 3 system, known as the doubly fed induction generator (DFIG), is the most common type of wind turbines. Although more expensive than types 1 and 2, it offers excellent operational and control features that make their integration with power grids easy and effective. Type 3 generator is a slip-ring machine similar to that of type 2. The generator is fed from the stator as well as the rotor (thus, doubly fed), as shown in the descriptive representation in Figure 10.1. The convention for the power flow of the DFIG is depicted in the figure. The mechanical power from the turbine enters the rotor and is converted into electrical (developed power). The rotor injection circuit also inserts power into the rotor. The stator power is the developed power plus the rotor power minus the losses. When the rotor power is negative, the flow is reversed: power is extracted from the rotor and delivered to the grid.

The main components of the DFIG are shown in Figure 10.2. The rotor of the generator is connected to the farm collection bus (FCB) via two voltage source converters: rotor-side converter (RSC) and grid-side converter (GSC). In the dc link between the two converters, a capacitor is placed to reduce the voltage ripples and store some energy.

The RSC injects voltage into the rotor of the generator at adjustable magnitude, frequency, and phase shift. Doing so, we can control several important variables such as

- Developed torque
- Speed of rotation
- Reactive power flow and power factor at the terminals of the stator
- Speed range of wind power generation
- Bidirectional power flow from rotor to dc link

The main objectives of the GSC are as follows:

- Keeps the voltage of the dc-link within a specified range
- Bidirectional power flow from dc link to grid
- Any flow toward the grid through the GSC is kept at the grid frequency
- Reactive power control

The industrial design of the DFIG varies based on the needed features and control strategies. One of the popular designs is the six pulse types shown in Figure 10.3. Each of these converters is dual-flow type; operates in dc/ac or ac/dc mode depending on desired flow of power. For example, if the flow of power is from the grid to the rotor, the GSC is in ac/dc mode and the RSC is in dc/ac mode. If the flow is reversed (from rotor to grid), the modes are reversed as well.

The transformer in the figure is used to step up the voltage V_{out} to the FCB voltage. It is also used to control the flow of reactive power from the GSC to the grid. In addition, it can provide filtering capability to the output current of the GSC.

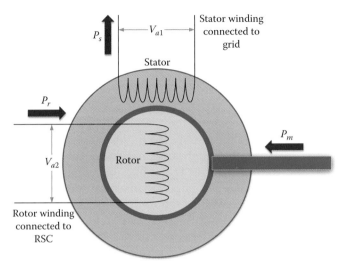

FIGURE 10.1
Concept of a DFIG.

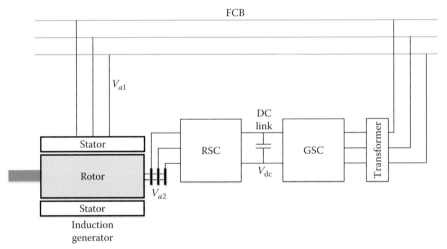

FIGURE 10.2
Main components of the DFIG.

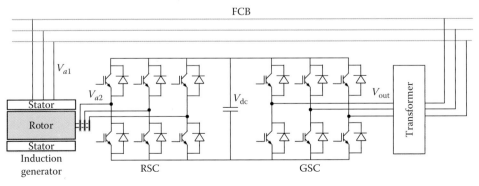

FIGURE 10.3
DFIG converters.

10.1 Equivalent Circuit

The equivalent circuit of the wound rotor machines is similar to that for the squirrel-cage type, but it includes the injected voltage as given in Chapter 6 and shown in Figure 10.4. V'_{a2} in the figure is the injected voltage referred to the stator windings. V_D is the voltage across the developed resistance r_d, which is defined in Chapter 8 as

$$\overline{V}_D = -r_d \overline{I}'_{a2} \tag{10.1}$$

where:

$$r_d = \frac{r'_2}{s}(1-s) \tag{10.2}$$

Because the actual injection (refereed to stator) is V'_{a2}, the component V'_{a2}/S can be parsed into two components, where one of them is the actual injection.

$$\frac{V'_{a2}}{s} = V'_{a2} + \frac{V'_{a2}}{s}(1-s) \tag{10.3}$$

The modified equivalent circuit that takes into account Equation 10.3 is shown in Figure 10.5. When we apply Thevinin's theorem to the stator circuit, the model is further modified, as shown in Figure 10.6, where

$$\overline{V}_{th} = jx_m \frac{\overline{V}_{a1}}{r_1 + j(x_1 + x_m)} \tag{10.4}$$

$$\overline{Z}_{th} = \frac{jx_m(r_1 + jx_1)}{r_1 + j(x_1 + x_m)} = r_{th} + jx_{th} \tag{10.5}$$

$$r_{eq} = r_{th} + r'_2$$
$$x_{eq} = x_{th} + x'_2 \tag{10.6}$$

Using the circuit in Figure 10.6, we can compute the rotor current $\overline{I}'_{a2}$

$$\overline{I}'_{a2} = \frac{\overline{V}_D + (V'_{a2}/s) - \overline{V}_{th}}{r_{eq} + jx_{eq}} \tag{10.7}$$

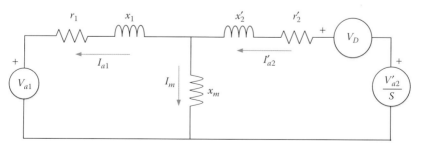

FIGURE 10.4
Model of the DFIG.

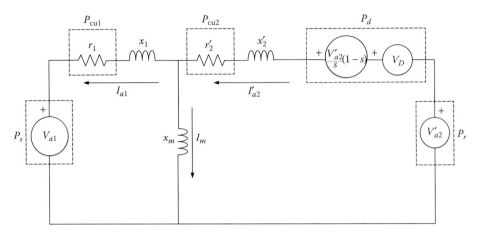

FIGURE 10.5
Power components in the DFIG.

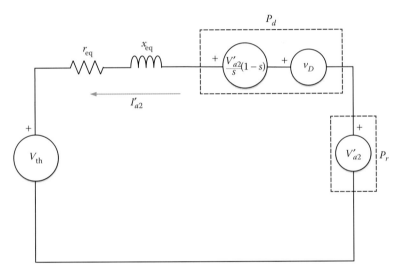

FIGURE 10.6
Thevinin's equivalent circuit for the DFIG.

Substituting $\bar{V}_D$ in Equation 10.1 into Equation 10.7 yields

$$\bar{I}'_{a2} = \frac{(V'_{a2}/s) - \bar{V}_{th}}{(r_{eq} + r_d) + jx_{eq}} \tag{10.8}$$

10.2 Simplified Model

If the resistance and inductive reactance of the stator winding are much smaller than the magnetizing inductive reactance x_m, we can simplify the model, as shown in Figure 10.7. In this case, we ignore the losses (real and reactive) of the stator, and Thevinin's voltage and impedance are

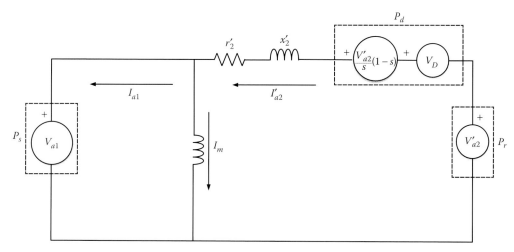

FIGURE 10.7
Simplified model of the DFIG.

$$\bar{V}_{th} = \bar{V}_{a1} \tag{10.9}$$

$$\bar{Z}_{th} = 0 \tag{10.10}$$

In this case, the rotor current in Equation 10.8 can be modified as

$$\bar{I}'_{a2} = \frac{(V'_{a2}/s) - \bar{V}_{a1}}{r'_d + r'_2 + jx'_2} \tag{10.11}$$

And the stator current is

$$\bar{I}_{a1} = \bar{I}'_{a2} - \frac{\bar{V}_{a1}}{jx_m} \tag{10.12}$$

10.3 Power Flow

The circuit in Figure 10.5 can be used to compute the various powers of the DFIG. For example, the electrical power coming from the rotor P_r is

$$P_r = 3I_{a2}V_{a2}\cos\theta_r = 3I'_{a2}V'_{a2}\cos\theta_r \tag{10.13}$$

where:
  θ_r is the angle between V'_{a2} and I'_{a2}

The stator power P_s delivered to the grid is

$$P_s = 3I_{a1}V_{a1}\cos\theta_s \tag{10.14}$$

where:

θ_s is the angle between V_{a1} and I_{a1}

The input power to the generator is equal to the output power plus all losses. The input power P_{in} is the mechanical power coming from the gearbox plus the rotor electrical power P_r coming through the RSC minus the rotational losses

$$P_{in} = P_m + P_r - P_{rotation} = P_d + P_r \tag{10.15}$$

where:

P_m is the mechanical power coming from the gearbox
$P_{rotation}$ is the rotational losses of the generator (often ignored)
P_d is the developed power (mechanical power converted into electrical)

The electrical losses of the generator are mainly due to the winding resistances. These are called "copper losses," P_{cu}.

$$P_{cu} = P_{cu1} + P_{cu2} \tag{10.16}$$

where:

P_{cu1} is the loss in the stator winding (stator copper loss)
P_{cu2} is the loss in the rotor winding (rotor copper loss)

Using the equivalent circuit in Figure 10.5, these losses can be computed as

$$P_{cu1} = 3r_1 I_{a1}^2$$
$$P_{cu2} = 3r_2' I_{a2}'^2 \tag{10.17}$$

Hence, the power balance equation is

$$P_s = P_d + P_r - P_{cu} \tag{10.18}$$

Using Equation 10.18 and the model in Figure 10.5, the developed power is then

$$P_d = 3V_D I_{a2}' \cos\theta_D + 3real\left[\frac{\overline{V}_{a2}'}{s}(1-s)\overline{I}_{a2}'\right] \tag{10.19}$$

$$P_d = -3r_d(I_{a2}')^2 + 3\frac{V_{a2}'}{s}(1-s)I_{a2}'\cos\theta_r \tag{10.20}$$

where:

θ_D is the angle between V_D and I_{a2}'
θ_r is the angle between V_{a2}' and I_{a2}'

The airgap power (P_g) is the power transferred to the stator by the flux in the airgap. It is equal to the developed power (P_d) plus the rotor power (P_r) minus the losses in the rotor winding (P_{cu}). This airgap power is also equal to the output power of the generator plus the stator copper loss.

$$P_g = P_d + P_r - P_{cu2} = P_s + P_{cu1} \tag{10.21}$$

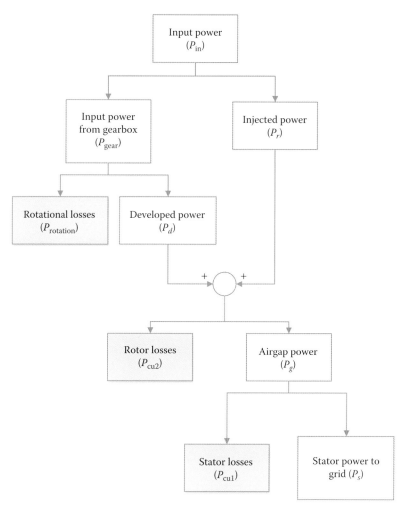

FIGURE 10.8
Power flow of wound rotor induction generator with rotor injection.

The power flow diagram representing all these powers is shown in Figure 10.8. The input powers are the mechanical power from the gearbox plus the injected rotor power from the RSC. Part of the mechanical power is lost due to friction and windage (rotational losses), and the rest is converted into electrical power (developed power). The developed power and the rotor injection power are added up to account for the total electrical power entering the rotor. Part of the total rotor power is lost in the form of rotor copper losses. The remaining power is transferred to the stator by the magnetic field in the airgap, thus called "airgap power." When the airgap power enters the stator, part of it is lost in the form of stator copper loss and the rest is the stator power (output power) delivered to the grid.

A power term that is commonly used in the industry is the *slip power* (P_{slip}), which is defined as the rotor power minus the rotor copper loss

$$P_{\text{slip}} \equiv P_r - P_{\text{cu2}} \tag{10.22}$$

Hence, using Equation 10.21, we can write the airgap power as

$$P_g = P_d + P_{slip} \tag{10.23}$$

As given in Chapters 6 and 8, the airgap and developed powers can be written in the mechanical terms

$$P_g = T_d \omega_s$$
$$P_d = T_d \omega_2 \tag{10.24}$$

Hence, the slip power is

$$P_{slip} = P_g - P_d = T_d \omega_s - T_d \omega_2 = T_d(\omega_s - \omega_2) = T_d s \omega_s = s P_g \tag{10.25}$$

Substituting Equation 10.25 into Equation 10.23 yields the relationship between the developed power and the airgap power

$$P_g = P_d + P_{slip} = P_d + s P_g \tag{10.26}$$

Hence,

$$P_d = P_g(1 - s) \tag{10.27}$$

Using the slip power, an alternative power flow diagram for the DFIG is shown in Figure 10.9.

Based on Equations 10.8 and 10.20, we can obtain the family of power–speed characteristics for various values of injected voltage that are shown in Figure 10.10. In this graph, the injected voltage is in line (in phase or 180° out of phase) with the stator voltage. From the characteristics, one can draw several conclusions, of which some of them are as follows:

- The machine can generate electricity even when the speed of the generator is less than the synchronous speed (sub-synchronous speed). This could not be done by types 1 or 2 systems as they must rotate above synchronous speed (super-synchronous speed) to generate electricity.
- The machine can operate at wide range of speed.
- Constant speed or constant power can be achieved by adjusting the injected voltage.

The DFIG can operate at positive or negative slips (sub- or super-synchronous speeds). The adopted notation for power flow in Figure 10.1 is for sub-synchronous speed; the slip is positive and the rotor power is positive as well. In this case, the flow of power is from the grid (though the converters) to the rotor. This is how the DFIG can generate electricity when the generator speed is below its synchronous speed.

In super-synchronous speed, the slip is negative and the rotor power flow is negative; that is, the power is from the rotor to the grid. In the case when the mechanical power from the turbine is higher than the rated power of the stator, the excess power goes to the grid through the rotor converters. Thus, increasing the total power-generation capability of the system.

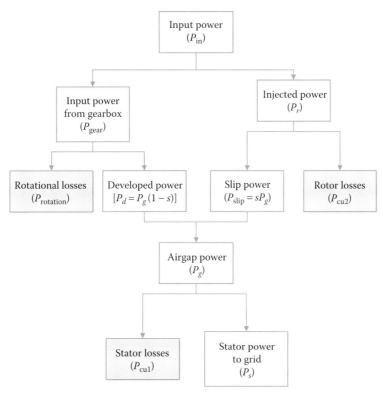

FIGURE 10.9
Alternative power flow diagram of the DFIG with rotor injection.

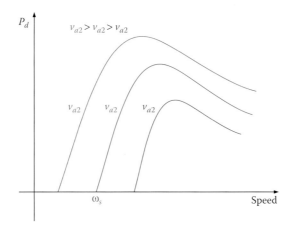

FIGURE 10.10
Power–speed characteristics of induction generator with rotor injection.

If s is maintained within ±30%, the power passing through the converter is about 30% of the stator rated power. Therefore, the DFIG's converter is designed to handle about one-third of the rated power. This way, the DFIG system can produce 30% more power as compared with types 1 or 2 systems, and it can operate at winder speed range.

EXAMPLE 10.1

A DFIG wind turbine has six-pole, 60 Hz three-phase, Y-connected induction generator. The terminal voltage of the generator is 690 V. The effective turns ratio of the generator is 2. The rotor parameters of the machine are as follows: $r_1 = r_2' = 5.0$ mΩ; $x_1 = x_2' = 150$ mΩ; and $x_m = 5$ Ω.

Assuming the stator voltage is the reference phasor, the injected voltage seen from the stator side is adjusted to $\bar{V}_{a2}' = 5\angle -120°$ V by a PWM technique. Compute the developed power, developed torque, rotor injected power, and stator power at 1230 rpm.

Solution:
The slip is

$$s = \frac{n_s - n}{n_s} = \frac{1200 - 1230}{1200} = -0.025$$

Thevinin's values in Equations 10.4 and 10.5 are

$$\bar{V}_{th} = jx_m \frac{\bar{V}_{a1}}{r_1 + j(x_1 + x_m)} = j5\frac{\left(690/\sqrt{3}\right)\sqrt{3}}{0.005 + j5.15} = 386.77\angle 0.056° \text{ V}$$

$$\bar{Z}_{th} = \frac{jx_m(r_1 + jx_1)}{r_1 + j(x_1 + x_m)} = \frac{j5\times(0.005 + j0.15)}{0.005 + j5.15} = 0.0047 + j0.1456 = 0.1457\angle 88.15°$$

Using Equation 10.8, we can compute the current $\bar{I}_{a2}'$

$$\bar{I}_{a2}' = \frac{(\bar{V}_{a2}'/s) - \bar{V}_{th}}{\left[r_{th} + (r_2' + r_d)\right] + jx_{eq}} = \frac{(5\angle -120°/-0.025) - 386.77\angle 0.056°}{\left[0.0047 + (0.005/-0.025)\right] + j0.2957} = 945\angle 25.48° \text{ A}$$

The developed power is computed from Equation 10.20

$$P_d = -3r_d(I_{a2}')^2 + 3\frac{V_{a2}'}{s}(1-s)I_{a2}'\cos\theta_r$$

$$P_d = -3\frac{0.005}{-0.025}\times 1.025\times 945^2 + 3\frac{5}{-0.025}\times 1.025\times 945\cos(-120 - 25.48) = 1.028 \text{ MW}$$

The developed torque is

$$T_d = \frac{P_d}{\omega_2} = 7.98 \text{ kNm}$$

Before computing the stator power, we need to calculate I_{a1}. The loop equation for the stator circuit is

$$\bar{V}_{a1} = -(r_1 + jx_1)\bar{I}_{a1} + jx_m\bar{I}_m = -(r_1 + jx_1)\bar{I}_{a1} + jx_m(\bar{I}_{a2}' - \bar{I}_{a1})$$

$$\bar{I}_{a1} = \frac{jx_m\bar{I}_{a2}' - \bar{V}_{a1}}{\left[r_1 + j(x_1 + x_m)\right]} = 953.3\angle 29.7° \text{ A}$$

The stator power can be computed using Equation 10.14

$$P_s = 3I_{a1}V_{a1}\cos\theta_s = \sqrt{3}\times 953.3\times 690\times\cos(29.7°) = 989.63 \text{ kW}$$

Stator losses

$$P_{cu1} = 3r_1 I_{a1}^2 = 13.395\,\text{kW}$$

Airgap power

$$P_g = P_s + P_{cu1} = 1.003\,\text{MW}$$

Rotor losses

$$P_{cu2} = 3r_2'(I_{a2}')^2 = 13.4\,\text{kW}$$

Slip power

$$P_{slip} = sP_g = -25.1\,\text{kW}$$

Injected rotor power from Equation 10.22

$$P_r = P_{slip} + P_{cu2} = -11.68\,\text{kW}$$

The negative sign indicates that P_r is going to the grid through the rotor. This is because the generator is operating at super-synchronous speed.

EXAMPLE 10.2

Repeat Example 10.1, but ignore all losses

Solution:
The slip is

$$s = \frac{n_s - n}{n_s} = \frac{1200 - 1230}{1200} = -0.025$$

To ignore the losses, we ignore the winding resistances. Hence, Thevinin's values in Equations 10.4 and 10.5 are

$$\bar{V}_{th} = jx_m \frac{\bar{V}_{a1}}{j(x_1 + x_m)} = 386.77\ \angle 0°\,\text{V}$$

$$\bar{Z}_{th} = \frac{jx_m(jx_1)}{j(x_1 + x_m)} = 0.1456\ \angle 90°$$

Using Equation 10.8, we can compute the current $\bar{I}_{a2}'$

$$\bar{I}_{a2}' = \frac{(\bar{V}_{a2}'/s) - \bar{V}_{th}}{r_d + jx_{eq}} = 931.24\ \angle -5.02°\,\text{A}$$

The error in current due to approximation is

$$\frac{945 - 931.24}{945} = 1.46\%$$

The developed power is computed from Equation 10.20

$$P_d = -3r_d'(I_{a2}')^2 + 3\frac{V_{a2}'}{s}(1-s)I_{a2}'\cos\theta_r = 997.42\,\text{kW}$$

The error in developed power due to approximation is

$$\frac{1028 - 997.42}{1028} = 2.97\%$$

The developed torque is

$$T_d = \frac{P_d}{\omega_2} = 7.74\,\text{kNm}$$

The error in developed torque due to approximation is the same as that for the developed power.

Before computing the stator power, we need to calculate I_{a1}. The loop equation for the stator circuit is

$$\bar{V}_{a1} = -jx_1\bar{I}_{a1} + jx_m\bar{I}_m = -jx_1\bar{I}_{a1} + jx_m(I_{a2}' - \bar{I}_{a1})$$

$$\bar{I}_{a1} = \frac{jx_mI_{a2}' - \bar{V}_{a1}}{j(x_1 + x_m)} = 938.39\,\angle 28.44°\,\text{A}$$

The stator power can be computed using Equation 10.14

$$P_s = 3I_{a1}V_{a1}\cos\theta_s = 986.1\,\text{kW}$$

The error in stator power due to approximation is

$$\frac{990.32 - 986.1}{990.32} = 0.04\%$$

Airgap power

$$P_g = P_s + P_{cu1} = 986.1 + 0 = 986.1\,\text{kW}$$

The error in the airgap power due to approximation is

$$\frac{1004 - 986.1}{1004} = 1.78\%$$

Slip power

$$P_{slip} = sP_g = -24.65\,\text{kW}$$

Injected rotor power from Equation 10.22

$$P_r = P_{slip} + P_{cu2} = -21.68 + 0 = -11.32$$

The error in the airgap power due to approximation is

$$\frac{-11.68 + 11.32}{-11.68} = 3.08\%$$

As seen from the above analyses, the approximate method produces very similar results if the electrical losses are insignificant.

EXAMPLE 10.3

A six-pole type 3 induction generator is spinning at 1350 rpm. The terminal voltage of the generator is 690 V. The parameters of the machine are $r_2' = 10$ mΩ; $x_2' = 100$ mΩ; and $x_m = 2$ Ω.

Ignore the stator winding impedance and compute the output power of the generator assuming that 30 V (referred to the stator) is injected into the rotor circuit and is in phase with the terminal voltage.

Solution:

The slip of the machine at 1350 rpm is

$$s = \frac{n_s - n}{n_s} = \frac{1200 - 1350}{1200} = -0.125$$

If we ignore the stator windings, the rotor current using Equation 10.7 is

$$\bar{I}_{a2}' = \frac{(\bar{V}_{a2}'/s) - \bar{V}_{a1}}{(r_2'/s) + jx_2'} = \frac{(30 \angle 0°/-0.125) - \left[(690/\sqrt{3})\angle 0°\right]}{(0.01/-0.125) + j0.1} = 4.985 \angle 51.34° \, \text{kA}$$

Because we are ignoring the stator windings, the voltage across the magnetizing branch is the same as the terminal voltage of the generator. Hence, the magnetizing current is

$$\bar{I}_m = \frac{\bar{V}_{a1}}{jx_m} = \frac{(690/\sqrt{3})\angle 0°}{j2} = 199.19 \angle -90° \, \text{A}$$

The current of the stator is

$$\bar{I}_{a1} = \bar{I}_{a2}' - \bar{I}_m = \bar{I}_{a2}' - \frac{\bar{V}_{a1}}{jx_m} = 4.985 \angle 51.34° - \frac{(690/\sqrt{3})\angle 0°}{j2} = 5.142 \angle 52.72° \, \text{kA}$$

The power delivered to the grid is

$$P_s = 3V_{a1}I_{a1}\cos\theta_1 = \sqrt{3} \times 690 \times 5.142 \times \cos 52.72° = 3.722 \, \text{MW}$$

Repeat the solution without the injected voltage. Can you draw a conclusion?

10.3.1 Apparent Power Flow through RSC

Figure 10.11 shows an induction generator with its RSC. The injected voltage through the RSC (V_{a2}) can be controlled by the PWM technique discussed in Chapter 5. The technique allows us to vary the magnitude, the phase shift, and the frequency of the injected voltage.

The components of the induction generator model in Figure 10.5 that are impacted by the reactive power are shown in Figure 10.12. These include all voltage sources and inductive reactances. Q_1 and Q_2 are the reactive power losses in the stator and rotor windings, respectively. Q_m is the reactive power consumed by the magnetizing branch represented by x_m. Q_s is the reactive power at the terminals of the generator, which is delivered to the grid. Because of the developed resistance r_d, the voltage source $\bar{V}_D$ and $\bar{I}_{a2}'$ are aligned as shown in Equation 10.1. Hence, $\bar{V}_D$ is not consuming or delivering any reactive power.

In the following analysis, we use the norm for the flow of reactive power shown in Figure 10.13. The reactive power is positive when it enters the windings.

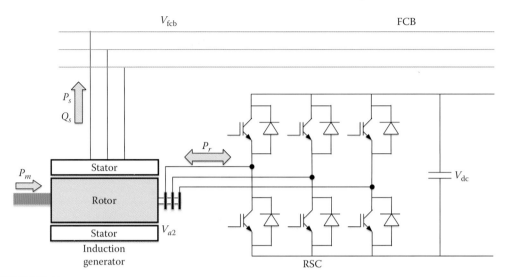

FIGURE 10.11
Power control through the rotor-side converter.

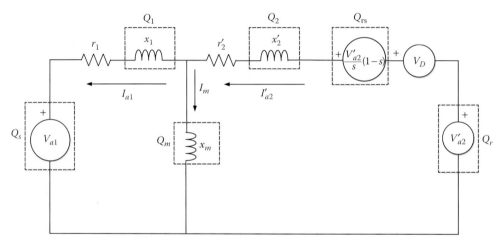

FIGURE 10.12
Reactive power of the DFIG.

The complex (apparent) power from the RSC ($\bar{S}_r$) is

$$\bar{S}_r = 3\bar{V}'_{a2}\bar{I}^*_{a2} = P_r + jQ_r$$

$$P_r = real\ (3\bar{V}'_{a2}\bar{I}^*_{a2})$$ (10.28)

$$Q_r = imag\ (3\bar{V}'_{a2}\bar{I}^*_{a2})$$

The term Q_{rs} is the reactive power produced in the rotor circuit due to the injected voltage and the slip of the machine.

$$Q_{rs} = imag\left[3\frac{\bar{V}'_{a2}}{s}(1-s)\bar{I}^*_{a2} \right]$$ (10.29)

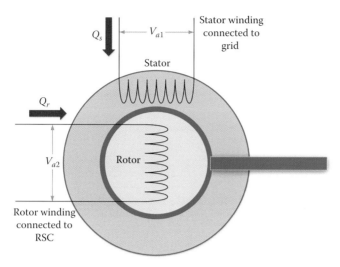

FIGURE 10.13
Reactive power flow in the DFIG.

The total reactive power in the rotor circuit as referred to the stator is

$$Q_{r-total} = Q_r + Q_{rs}$$ (10.30)

Substituting Q_r in Equation 10.28 and Q_{rs} in Equation 10.29 into Equation 10.30 yields

$$Q_{r-total} = imag\left(3\frac{\overline{V}'_{a2}}{s}\overline{I}'^*_{a2}\right) = \frac{Q_r}{s}$$ (10.31)

Equation 10.31 shows that the reactive power Q_r injected by the RSC is amplified by a factor of $1/s$ when seen by the stator. This is a great feature that allows us to effectively use the RSC in reactive power control.

The balance of reactive power is when the reactive powers produced by all sources is equal to the reactive power consumed by all inductive reactances.

$$Q_s + Q_{r-total} = Q_1 + Q_2 + Q_m$$

$$Q_s + \frac{Q_r}{s} = Q_1 + Q_2 + Q_m$$ (10.32)

EXAMPLE 10.4

A 690 V, 12-pole DFIG system has a generator with the following parameters: $r'_2 = 10$ mΩ; $x_1 = x'_2 = 100$ mΩ; and $x_m = 5$ Ω. The machine operates at 660 rpm. If the injected voltage referred to the stator is $40 \angle -150°$ V, compute the reactive power delivered to the grid.

Solution:
At 660 rpm, the slip of the machine is

$$s = \frac{n_s - n}{n_s} = \frac{600 - 660}{600} = -0.1$$

To compute the powers, we need to compute the currents. Using Equations 10.4 and 10.5 to compute Thevinin's equivalent circuit, we get

$$\bar{V}_{th} = jx_m \frac{\bar{V}_{a1}}{r_1 + j(x_1 + x_m)} = j5\frac{690/\sqrt{3}}{0.01 + j(5.1)} = 390.56 + j0.766 \text{ V}$$

$$\bar{Z}_{th} = \frac{jx_m(r_1 + jx_1)}{r_1 + j(x_1 + x_m)} = 0.0096 + j0.098$$

Use Equation 10.8 to compute the rotor current referred to stator

$$\bar{I}'_{a2} = \frac{(\bar{V}'_{a2}/s) - \bar{V}_{th}}{(r_{eq} + r_d) + jx_{eq}} = 916.7 - j195.46 \text{ A}$$

Using the loop equation for the rotor circuit, we can compute the voltage across the magnetizing branch

$$\bar{V}_m = \frac{\bar{V}'_{a2}}{s} - \bar{I}'_{a2}\left[(r'_2 + r_d) + jx'_2\right] = 418.54 + j88.78 \text{ V}$$

Hence,

$$\bar{I}_{a1} = \bar{I}'_{a2} - \frac{\bar{V}_m}{jx_m} = 899 - j111.75 \text{ A}$$

Now, let us compute the various reactive powers. We can start with the reactive power consumptions

$$Q_2 = 3x'_2 I'^2_{a2} = 263.58 \text{ kVAr}$$

$$Q_1 = 3x_1 I^2_{a1} = 246.2 \text{ kVAr}$$

$$Q_m = 3x_m I^2_m = 109.83 \text{ kVAr}$$

Compute the reactive power produced by the RSC using Equation 10.28

$$Q_r = imag\,(3\bar{V}'_{a2}\bar{I}'^*_{a2}) = -75.32 \text{ kVAr}$$

The total reactive power of the rotor can be computed using Equation 10.31

$$Q_{r-total} = \frac{Q_r}{s} = 753.2 \text{ kVAr}$$

From Equation 10.32, we can compute the reactive power at the terminals of the generator

$$Q_s = Q_1 + Q_2 + Q_m - Q_{r-total} = -133.59 \text{ kVAr}$$

The negative sign of Q_s indicates that the flow of reactive power is from the generator to the grid.

Repeat the calculations when the injected voltage referred to the stator is $40 \angle +150°$V. Can you draw a conclusion?

10.3.2 Apparent Power Flow through GSC

The GSC can also control the apparent power flow to the grid. Consider the system from the dc link to the grid shown in Figure 10.14. In this system, the magnitude, phase shift, and frequency of the output voltage of the GSC (V_{out}) are controlled by a PWM technique. The frequency of V_{out} is adjusted to match the frequency of the grid. The magnitude and phase shift of V_{out} are the two control variables that allows us to control the flow of power to the grid. Between the output of the GSC and the farm collection bus (FCB), a transformer is installed.

The one-line diagram of the system in Figure 10.14 is shown in Figure 10.15. In other systems, an inductor is used instead of the transformer. The one-line diagram with power flows is shown in Figure 10.16. In the following analysis, we assume that the real power losses of the transformer are insignificant. The inductive reactance (x) in the figure is the equivalent inductive reactance of the transformer windings as referred to the FCB side of the transformer. V'_{out} is the GSC output refereed to the FCB side of the transformer.

$$V'_{out} = V_{out} \frac{N_2}{N_1} \tag{10.33}$$

where:
N_1 is the number of turns of the transformer's winding connected to the GSC
N_2 is the number of turns of the transformer's winding connected to the FCB

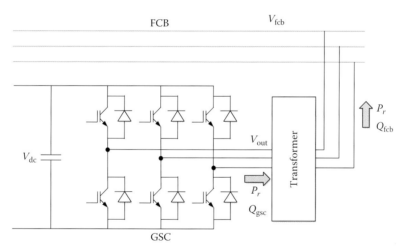

FIGURE 10.14
Reactive power control through a grid-side converter.

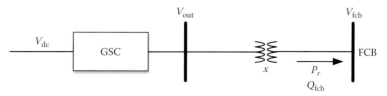

FIGURE 10.15
One-line diagram of the system in Figure 10.14.

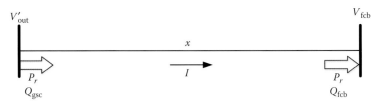

FIGURE 10.16
Equivalent line diagram of the system in Figure 10.15.

Let us assume a super-synchronous speed operation. During the steady state, the real power output of the GSC is the rotor power P_r. However, the reactive power at the GSC and FCB are different because of the reactive power consumed by the inductive reactance of the transformer.

To compute all various powers, we need to compute the current going to the FCB from GSC.

$$\bar{I} = \frac{\bar{V}'_{out} - \bar{V}_{fcb}}{\bar{x}} \tag{10.34}$$

If we take $\bar{V}_{fcb}$ as a reference

$$\bar{I} = \frac{V'_{out}(\cos\delta + j\sin\delta) - V_{fcb}}{jx} = \frac{V'_{out}}{x}\sin\delta + j\frac{V_{fcb} - V'_{out}\cos\delta}{x} \tag{10.35}$$

where:
δ is the angle of $\bar{V}'_{out}$ with respect to $\bar{V}_{fcb}$, which is also known as the "power angle"

The complex (apparent) power delivered to the FCB is

$$\bar{S}_{fcb} = 3\bar{V}_{fcb}\,\bar{I}^* = P_r + jQ_{fcb} \tag{10.36}$$

Substituting the value of $\bar{I}$ in Equation 10.35 into Equation 10.36 yields the real and reactive power delivered to the grid through the GSC.

$$\bar{S}_{fcb} = P_r + jQ_{fcb} = 3V_{fcb}\left[\left(\frac{V'_{out}}{x}\sin\delta\right) - \left(j\frac{V_{fcb} - V'_{out}\cos\delta}{x}\right)\right] \tag{10.37}$$

Hence,

$$P_r = 3\frac{\bar{V}'_{out}V_{fcb}}{x}\sin\delta \tag{10.38}$$

$$Q_{fcb} = 3\frac{V_{fcb}}{x}(V'_{out}\cos\delta - V_{fcb}) \tag{10.39}$$

EXAMPLE 10.5

For the system in Example 10.4, assume the inductive reactance of the transformer is 2 Ω. Compute the magnitude and angle of $\bar{V}'_{out}$ that would deliver 20.77 kW and 6 kVAr to the FCB.

Solution:
Use Equation 10.38 and 10.39 to compute $\bar{V}'_{out}$

$$P_r = \frac{\bar{V}'_{out} V_{fcb}}{x} \sin\delta$$

$$\frac{20.77 \times 10^3}{3} = \frac{690/\sqrt{3}}{2} \bar{V}'_{out} \sin\delta$$

$$V'_{out} \sin\delta = 34.758 \text{ V}$$

For the reactive power

$$Q_{fcb} = \frac{V_{fcb}}{x}(\bar{V}'_{out} \cos\delta - V_{fcb})$$

$$\frac{6000}{3} = \frac{690/\sqrt{3}}{2}\left(V'_{out} \cos\delta - \frac{690}{\sqrt{3}}\right)$$

$$V'_{out} \cos\delta = 408.41 \text{ V}$$

Hence, the angle of the injected voltage is

$$\delta = \tan^{-1}\left(\frac{V'_{out} \sin\delta}{\bar{V}'_{out} \cos\delta}\right) = \tan^{-1}\left(\frac{34.758}{408.41}\right) = 4.86°$$

The magnitude of the phase value of $\bar{V}'_{out}$ is

$$V'_{out} = \frac{34.758}{\sin\delta} = 409.88 \text{ V}$$

10.4 Speed Control

An important feature that gives type 3 system a great deal of flexibility is its ability to control the speed of rotation, track the maximum coefficient of performance, and generate electricity at a relatively wider speed range, including sub-synchronous speeds. To explain these control actions, let us ignore the rotational losses and write the dynamic equations of the rotor, as given in Chapter 6.

$$T_m - T_d = 2H\frac{d\omega_2}{dt} \tag{10.40}$$

where:
T_m is the input mechanical torque from the gearbox
T_d is the developed electrical torque of the generator, $T_d = (P_d/\omega_2)$
H is the inertia time constant of the rotating mass (blades, generator, gearbox, etc.)
$d\omega_2/dt$ is the acceleration of the generator speed

Equation 10.20 shows that the electric power (or torque) is controlled by the current. The current, in return is controlled by the injected voltage, as given in Equation 10.8. Using these equations, we can explain how the speed of the rotation is controlled to achieve objectives such as constant speed operation or tracking the maximum coefficient of performance.

- *Constant speed operation:* Any change in wind speed changes the lift force, and consequently increasing the mechanical torque from the gearbox. Assume our objective is to keep the rotation speed of the turbine constant even when wind speed changes. In this case, the injected voltage is adjusted to change the electrical (developed) torque to match the change in the mechanical torque. Thus, the acceleration of the rotor is kept at zero, and the speed is maintained constant.
- *Tracking maximum C_p:* To track the maximum coefficient of performance, the tip-speed ratio (TSR) must be controlled. If wind speed changes, the speed of the generator must change to achieve the optimum TSR. This is done by adjusting the injected voltage to produce a temporary change in the electrical torque until the desired speed is achieved. After reaching the new speed, the injected voltage is adjusted again to match the electrical torque to the mechanical torque, thus operating at the new speed.

Keep in mind that controlling the pitch angle can also achieve speed control. However, adjusting the pitch angle is much slower than adjusting the injected voltage. This is due to the heavy inertia in the system and the limitation imposed on the speed of pitch angle actuation due to blade wobbling.

EXAMPLE 10.6

A DFIG system has a six-pole, 60 Hz, Y-connected induction generator. The terminal voltage of the generator is 690 V. The effective turns ratio of the generator is 5. The rotor parameters of the machine are $r_2' = 10$ mΩ; $x_1 = x_2' = 100$ mΩ; and $x_m = 5$ Ω.

Without injection, use the approximate method and compute the developed power at 1230 rpm.

If the speed of the generator increases to 1260 rpm, compute the injected voltage that maintains the developed power constant. Assume the injected voltage is 180° out of phase with the stator voltage.

Solution:
Without injection, the slip is

$$s = \frac{n_s - n}{n_s} = \frac{1200 - 1230}{1200} = -0.025$$

$$r_d = \frac{r_2'}{s}(1-s) = \frac{0.01}{-0.025}(1.025) = -0.41 \ \Omega$$

Because $x_1 \ll x_m$, the current in Equation 10.8 without injection can be approximated as

$$I_{a2}' = \frac{V_{a1}}{\sqrt{r_d^2 + (x_1 + x_2')}}$$

The developed power can be obtained from Equation 10.20 without injection

$$P_d = -3r_d(I'_{a2})^2 = -3r_d \frac{V_{a1}^2}{r_d^2 + (x_1 + x'_2)^2} = 0.41 \frac{690^2}{(-0.41)^2 + (0.2)^2} = 938.015\,\text{kW}$$

With injection, the slip is

$$s = \frac{n_s - n}{n_s} = \frac{1200 - 1360}{1200} = -0.05$$

The current of the generator with injection is given in Equation 10.8

$$I'_{a2} = \frac{(-V'_{a2}/s) - V_{a1}}{\sqrt{r_d^2 + (x_1 + x'_2)^2}}$$

where:

$$r_d = \frac{r'_2}{s}(1-s) = \frac{0.01}{-0.05}(1.05) = -0.21\,\Omega$$

Hence,

$$I'_{a2} = \frac{(-V'_{a2}/s) - V_{a1}}{\sqrt{r_d^2 + (x_1 + x'_2)^2}} = \frac{\left(-V'_{a2}/-0.05\right) - \left(690/\sqrt{3}\right)}{\sqrt{(-0.21)^2 + (0.2)^2}} = 68.97 \times V'_{a2} - 1373.7$$

The developed power with rotor injection can be obtained from Equation 10.20

$$P_d = -3r_d(I'_{a2})^2 + 3\frac{V'_{a2}}{s}(1-s)I'_{a2}\cos\theta_r$$

Because the injected voltage is $180°$ out of phase with the stator voltage, $\theta_r = 180°$.

$$P_d = -3r_d(I'_{a2})^2 - 3\frac{V'_{a2}}{s}(1-s)I'_{a2}$$

$$P_d = 0.63(I'_{a2})^2 - 3\frac{V'_{a2}}{-0.05}(1.05)I'_{a2} = 0.63(I'_{a2})^2 + 63V'_{a2}I'_{a2}$$

The current and developed power equations has two unknowns: V'_{a2} and I'_{a2}. Substituting the current into the above equation yields

$$P_d = 938.015 \times 10^3 = 0.63(68.97 \times V'_{a2} - 1373.7)^2 + 63V'_{a2}(68.97 \times V'_{a2} - 1373.7)$$

$$0 = 9.102 \times (V'_{a2})^2 + 102.807V'_{a2} - 2825.06$$

Hence, the phase value of V'_{a2} is

$$V'_{a2} = 12.85\,\text{V}$$

The phase value of the actual injected voltage V_{a2} is

$$V_{a2} = V'_{a2}\frac{N_2}{N_1} = 12.85\frac{1}{5} = 2.57\,\text{V}$$

EXAMPLE 10.7

A type 3 wind turbine has a six-pole, 60 Hz three-phase, Y-connected induction generator. Without injection and at a given wind condition, the power delivered to the grid is the rated power of the stator, which is 2 MW. When wind speed increases, so that the power from the gearbox is 2.5 MW, an injected voltage is adjusted to maintain the power through the stator at its rated value. Compute the power to the grid through the converter and the speed of the generator.

Solution:

If we ignore the losses, the power before wind speed increase is

$$P_{m1} = P_{d1} = P_s = 2.0\,\mathrm{MW}$$

When wind speed increases,

$$P_{m2} = P_{d2} = 2.5\,\mathrm{MW}$$

Hence, the power through the rotor is

$$P_r = P_{m2} - P_s = 0.5\,\mathrm{MW}$$

The slip of the machine can be obtained from Equations 10.22 through 10.25. If losses are ignored, we get

$$P_{\mathrm{slip}} = P_g - P_d = P_r = s P_g$$

Because we are ignoring the losses, the slip is

$$s \approx \frac{P_r}{P_g} = \frac{P_r}{P_s} = \frac{-0.5}{2} = -0.25$$

The negative sign of the value of P_r is introduced as the power flows from the rotor to the grid.

$$n = n_s(1-s) = 1200(1+0.25) = 1500\,\mathrm{rpm}$$

10.5 Protection of Type 3 Systems

A significant disadvantage of wind turbines is their vulnerability to grid disturbances. The DFIG, in particular, is more vulnerable than types 1 and 2 because of the presence of power electronic converters (RSC and GSC). The devices used in these converters are highly reliable with long life time, as long as they are not operating outside their limits such as the following:

Steady-state limit: The device can operate continuously without being damaged as long as the current and voltage are below the device limits.

Junction temperature: Losses inside solid-state devices are due to impurities of their material as well as the operating conditions of their circuits. These losses mainly occur during continuous operation as well as when the device toggles between the open and closed states. These electrical losses are converted into thermal energy that heats the junctions of the devices. The heat can be tolerated up to a critical level (125°C is a typical value). To help dissipate the junction temperature, a cooling method such as heat sinks, fluid, or gas cooling is used to transfer the heat to the surrounding environment. If the rate of heat accumulation is more than the rate of cooling, the device will permanently fail.

Surge current: It is the absolute maximum of the nonrepetitive impulse current.

Critical rate of rise of current (or maximum di/dt): Any solid-state device is damaged if the disturbance in the grid causes the current through the device to change rapidly (high *di/dt*), even if the current is below the surge limit of the device.

Critical rate of rise of voltage (or maximum dv/dt): When the rate of change in voltage across a device (*dv/dt*) exceeds its limit, the device is forced to close. This is a form of false triggering and could lead to excessive current or excessive *di/dt*.

In addition to power electronic failures, grid disturbances could lead to energy imbalance for the generator that could lead of severe mechanical stress that could damage the gearbox, generator, or even the structure of the turbine itself.

To safeguard the turbine during grid disturbances, two types of protections are employed:

Electrical protection: The converters are susceptible to overcurrent and overvoltage that could occur during grid disturbances such as faults. These events can cause current and voltage surges in the windings of the machine. Because the RSC is connected to the rotor of the generator, these surges could reach the RSC and could fail its power electronic devices.

Electromechanical protection: Because grid faults depress the voltage, the output power of the generator P_s is reduced during faults. Because the mechanical power of the machine cannot be reduced instantly, there will be a period of time when the mechanical power is higher than the delivered electrical power. This difference in power causes the turbine to accelerate and it may reach a level that could damage its blades as well as various other mechanical parts. To protect the turbine, the balance of power must be restored quickly, and the excess kinetic energy in its rotating mass must be dissipated before the speed reaches the upper limit of the turbine.

10.5.1 Electrical Protection

To protect the converters from transient overcurrent and overvoltage during system disturbances, several methods are employed. The most common ones are the crowbar and the chopper systems. The crowbar system is installed between the rotor and the RSC. The chopper system is installed in the dc link bus between the RSC and the GSC. Most type 3 systems employ at least one of these two systems.

10.5.1.1 Crowbar System

The crowbar is a three-phase resistance bank connected to the rotor windings of the DFIG, as shown in Figure 10.17. When the absolute value of the rotor current, or the dc link voltage, reaches an upper threshold during grid disturbances, the crowbar circuit is activated and the triggering of the RSC transistors are blocked. This way, the damaging high current of the rotor does not reach the converter, and the magnitude of the current is regulated by the switching circuit of the crowbar resistance. The machine in this case behaves like an induction machine with dynamic braking resistance. In most designs, the crowbar is activated for up to 2 seconds, which is enough time for circuit breakers to clear most grid faults.

One obvious question is why not just short the windings of the rotor during faults. After all, this will ensure that no current would flow through RSC. The problem is that the generator has small winding resistances that limit the current and provide poor damping for current transients. When the crowbar resistance is added to the rotor circuit, it reduces the current and increases the damping. Two major objectives must be considered in the design of the crowbar resistance R_{cb}.

1. *Provide damping:* R_{cb} should be high enough to limit and damp the current transients. Current damping is proportional to the value of the resistance.

2. *Limit voltage on RSC:* R_{cb} should be low enough to reduce the voltage across the RSC.

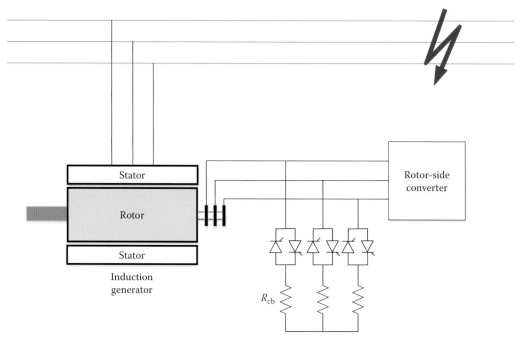

FIGURE 10.17
Crowbar system.

For the first objective, we can study the impact of the crowbar resistance on system oscillation. To do that, let us examine the dynamic model of the induction machine developed in Chapter 6 (Equation 6.107).

$$\frac{d}{dt}\begin{bmatrix} i_{d1} \\ i_{q1} \\ i_{d2} \\ i_{q2} \end{bmatrix} = \frac{1}{L}\begin{bmatrix} r_1 L_{22} & -L_{11}L_{22}\omega + L_{12}^2 s\omega & -r_2 L_{12} & -L_{12}L_{22}\omega_2 \\ L_{11}L_{22}\omega - L_{12}^2 s\omega & r_1 L_{22} & L_{12}L_{22}\omega_2 & -r_2 L_{12} \\ -r_1 L_{12} & L_{11}L_{12}\omega_2 & r_2 L_{11} & L_{12}^2\omega - L_{11}L_{22}s\omega \\ -L_{11}L_{12}\omega_2 & -r_1 L_{12} & -L_{12}^2\omega + L_{11}L_{22}s\omega & r_2 L_{11} \end{bmatrix}\begin{bmatrix} i_{d1} \\ i_{q1} \\ i_{d2} \\ i_{q2} \end{bmatrix}$$

$$+ \frac{1}{L}\begin{bmatrix} -L_{22} & 0 & L_{12} & 0 \\ 0 & -L_{22} & 0 & L_{12} \\ L_{12} & 0 & -L_{11} & 0 \\ 0 & L_{12} & 0 & -L_{11} \end{bmatrix}\begin{bmatrix} v_{d1} \\ v_{q1} \\ v_{d2} \\ v_{q2} \end{bmatrix} \tag{10.41}$$

The model can be written in the state space form of

$$\frac{d}{dt}\begin{bmatrix} i_{d1} \\ i_{q1} \\ i_{d2} \\ i_{q2} \end{bmatrix} = \mathbf{A}_\omega \begin{bmatrix} i_{d1} \\ i_{q1} \\ i_{d2} \\ i_{q2} \end{bmatrix} + \mathbf{B}\begin{bmatrix} v_{d1} \\ v_{q1} \\ v_{d2} \\ v_{q2} \end{bmatrix} \tag{10.42}$$

The eigenvalues of $\mathbf{A}_\omega$ can tell us the amount of damping in the system as well as the oscillation frequency. With crowbar system, the crowbar resistance augments the existing rotor resistance. The total resistance in the rotor is the resistance of the rotor windings (r_2) plus the resistance of the crowbar (R_{cb}). Equation 10.41 can be rewritten for this case as

$$\frac{d}{dt}\begin{bmatrix} i_{d1} \\ i_{q1} \\ i_{d2} \\ i_{q2} \end{bmatrix} = \frac{1}{L}\begin{bmatrix} r_1 L_{22} & -L_{11}L_{22}\omega + L_{12}^2 s\omega & -(r_2 + R_{cb})L_{12} & -L_{12}L_{22}\omega_2 \\ L_{11}L_{22}\omega - L_{12}^2 s\omega & r_1 L_{22} & L_{12}L_{22}\omega_2 & -(r_2 + R_{cb})L_{12} \\ -r_1 L_{12} & L_{11}L_{12}\omega_2 & (r_2 + R_{cb})L_{11} & L_{12}^2\omega - L_{11}L_{22}s\omega \\ -L_{11}L_{12}\omega_2 & -r_1 L_{12} & -L_{12}^2\omega + L_{11}L_{22}s\omega & (r_2 + R_{cb})L_{11} \end{bmatrix}\begin{bmatrix} i_{d1} \\ i_{q1} \\ i_{d2} \\ i_{q2} \end{bmatrix}$$

$$+ \frac{1}{L}\begin{bmatrix} -L_{22} & 0 & L_{12} & 0 \\ 0 & -L_{22} & 0 & L_{12} \\ L_{12} & 0 & -L_{11} & 0 \\ 0 & L_{12} & 0 & -L_{11} \end{bmatrix}\begin{bmatrix} v_{d1} \\ v_{q1} \\ v_{d2} \\ v_{q2} \end{bmatrix} \tag{10.43}$$

$\mathbf{A}_\omega$ in Equation 10.42 is changed to a new matrix $\hat{\mathbf{A}}_\omega$ due to the insertion of the crowbar resistance. The model can now be written in the state space form of

$$\frac{d}{dt}\begin{bmatrix} i_{d1} \\ i_{q1} \\ i_{d2} \\ i_{q2} \end{bmatrix} = \hat{\mathbf{A}}_\omega \begin{bmatrix} i_{d1} \\ i_{q1} \\ i_{d2} \\ i_{q2} \end{bmatrix} + \mathbf{B}\begin{bmatrix} v_{d1} \\ v_{q1} \\ v_{d2} \\ v_{q2} \end{bmatrix} \tag{10.44}$$

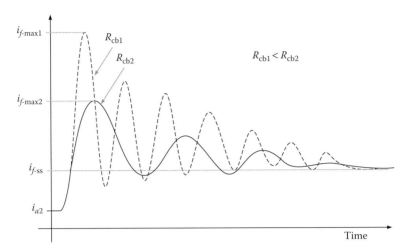

FIGURE 10.18
Current damping for different values of crowbar resistance.

By adding the crowbar resistance, the negative real components of the eigenvalues of $\hat{\mathbf{A}}_\omega$ is higher than that for $\mathbf{A}_\omega$. Hence, the system damping is enhanced. Figure 10.18 shows a typical transient current of induction generator under fault condition. Before the fault, the current of the rotor is i_{a2}. After the fault, the transient current reaches a peak value of $i_{f\text{-max}}$ and oscillates until it is damped out to a steady-state fault current $i_{f\text{-ss}}$. The resistances and inductances of the generator determine the magnitude of $i_{f\text{-max}}$ and $i_{f\text{-ss}}$, as well as the damping coefficient and oscillation frequency. The transient currents for two crowbar resistances are shown in the figure. With the smaller crowbar resistance, the maximum transient current $i_{f\text{-max}}$ is larger and is damped out after a longer time.

For the second objective, if R_{cb} is too large, the transient voltage across the crowbar terminals may become higher than the voltage of the dc link. In this case, excessive transient currents could flow through the anti-parallel diodes of the RSC shown in Figure 10.3 even when the transistors are not triggered. This could overcharge the capacitor of the dc link causing it to fail. It could also cause the power electronic devices connected to the dc link to fail due to overcurrent, overvoltage, high di/dt, or high dv/dt.

The crowbar resistance should be selected to reduce $i_{f\text{-max}}$ and damp its oscillations. In addition, the voltage across the crowbar resistance at $i_{f\text{-max}}$ must be lower than the voltage of the dc link,

$$R_{cb}\, i_{f\text{-max}} < V_{dc} \qquad (10.45)$$

In this case, the path of the transient current from the rotor to the dc bus through the anti-parallel diodes of the RSC is eliminated.

One drawback of the crowbar protection is the lack of reactive power control during faults. Because the RSC is disabled when the crowbar protection is activated, the reactive power of the generator will have to come from the grid. This is a violation of some grid codes as it could exacerbate the deterioration of system voltage. In countries with reactive current injection requirement during fault, the crowbar system must actuate for a short time in order to enable some reactive power injection by the RSC and GSC.

10.5.1.2 Chopper System

If the rotor current of the generator during grid faults reaches the dc link, it could elevate the dc bus voltage due to the acquired extra energy by the capacitor. The elevation of the dc bus voltage could damage the power electronic converters. One solution for this problem is to use the crowbar system to absorb the energy in the resistance of the crowbar. The chopper is another solution that can be used on its own or in conjunction with the crowbar circuit. The chopper circuit is located in the dc link and is in parallel with the capacitor, as shown in Figure 10.19. It is designed to dissipate the extra energy in its resistance R_{ch} to maintain the voltage of the dc bus to a desired value. The duty ratio of the transistor of the chopper circuit is controlled to allow R_{ch} to consume the needed amount of energy. By this method, the voltage of the dc link can be kept to a safe range.

The drawback of this method is that the high current during fault must pass through the RSC to reach the chopper. Hence, the RSC components must be designed to handle these high currents.

Sizing the resistance of the chopper (R_{ch}) can be made using approximate calculations. The first step is to compute the energy stored in the capacitor during steady-state operation.

$$E_{co} = \frac{1}{2}CV_{dco}^2 \qquad (10.46)$$

where:
E_{co} is the stored energy in the capacitor during steady-state operation
V_{dco} is the steady-state voltage of the dc link
C is the capacitance of the capacitor

When the voltage of the dc link increases to V_{dc} due to surges in rotor currents, the new energy of the capacitor is

$$E_c = \frac{1}{2}CV_{dc}^2 \qquad (10.47)$$

where:
E_c is the stored energy in the capacitor after the current surge
V_{dc} is the new voltage of the dc link, $V_{dc} > V_{dco}$

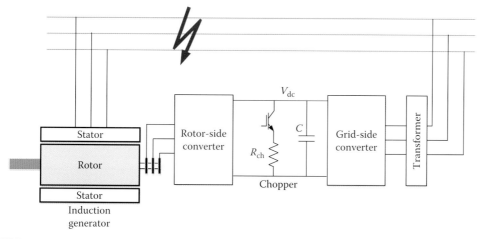

FIGURE 10.19
Chopper system.

The extra energy acquired by the capacitor due to the current surge is

$$\Delta E_c = E_c - E_{co} = \frac{1}{2}C(V_{dc}^2 - V_{dco}^2) \tag{10.48}$$

The power consumed by the resistance of the chopper is

$$P_R = \frac{V_R^2}{R_{ch}} \tag{10.49}$$

V_R is the voltage across the chopper resistance, which is

$$V_R = kV_{dc} \tag{10.50}$$

$$k = \frac{t_{on}}{\tau} \tag{10.51}$$

where:
    k is the duty ratio of the chopper's transistor
    t_{on} is the on-time of the transistor in one period
    τ is the period of the switching cycle of the transistor

Replacing V_R in Equation 10.49 by that in Equation 10.50 yields

$$P_R = \frac{V_{dc}^2}{R_{ch}}k^2 \tag{10.52}$$

The energy consumed by the chopper resistance is

$$E_R = P_R t = \frac{V_{dc}^2}{R_{ch}}k^2 t \tag{10.53}$$

where:
    t is the desired time to dissipate the extra energy

To keep the voltage at or near the steady-state value, the chopper resistance must consume the extra energy ΔE_c. Hence,

$$E_R = \Delta E_c$$
$$\frac{V_{dc}^2}{R_{ch}}k^2 t = \frac{1}{2}C(V_{dc}^2 - V_{dco}^2) \tag{10.54}$$

The duty ratio of the chopper can be computed from Equation 10.54

$$k = \sqrt{\frac{C\,R_{ch}}{2t}\left[1 - \left(\frac{V_{dco}}{V_{dc}}\right)^2\right]} \tag{10.55}$$

The equation shows that the duty ratio depends on three parameters: the ohmic value of the chopper resistance (R_{ch}), the allowed increase in voltage in the dc bus (V_{dc}), and the time to dissipate the extra energy (t).

EXAMPLE 10.8

The rated voltage of the dc bus of the DFIG turbine is 1000 V. The dc bus capacitor is 100 μF and the chopper resistance is 2 Ω. The allowed increase in dc bus voltage is 10%. Compute the duty ratio that dissipate the extra energy in 100 μs. Repeat the calculations for a voltage increase of 20%.

Solution:
For 10% voltage increase, the duty ratio of the chopper's transistor can be computed using Equation 10.55.

$$k = \sqrt{\frac{C\,R_{ch}}{2t}\left[1 - \left(\frac{V_{dco}}{V_{dc}}\right)^2\right]} = \sqrt{\frac{10^{-4} \times 2}{2 \times 10^{-4}}\left[1 - \left(\frac{1000}{1100}\right)^2\right]} = 0.416$$

For 20% voltage increase

$$k = \sqrt{\frac{C\,R_{ch}}{2t}\left[1 - \left(\frac{V_{dco}}{V_{dc}}\right)^2\right]} = \sqrt{\frac{10^{-4} \times 2}{2 \times 10^{-4}}\left[1 - \left(\frac{1000}{1200}\right)^2\right]} = 0.553$$

The switching frequency for the transistor can be arbitrary selected; a value between 5 and 15 kHz is within power electronic switching capabilities

10.5.2 Electromechanical Protection

The voltage of the power grid is suppressed during faults, and the largest reduction occurs near the fault. The depression in voltage reduces the electrical power output of the generator for as long as the fault lasts. From the turbine viewpoint, the reduction of its electrical power output must be matched by a quick reduction in the mechanical power captured by the blades. Otherwise, the turbine will accelerate and the speed of the blades could reach a destructive level. During the period from the initiation of the fault and until the blades are pitched to the correct angle, the mechanical power going into the generator P_m is higher than the electrical power output of the generator P_{out}. If we ignore the losses of the windings, the power imbalance of the generator is

$$P_m > P_{out} = (P_s + P_r) \tag{10.56}$$

From the energy viewpoint, we can write the above equation as

$$\int_0^t P_m dt > \int_0^t P_{out} dt \tag{10.57}$$

$$E_m > E_{out} = (E_s + E_r)$$

where:
  t is the time during which the voltage is depressed
  E_m is the energy provided by the turbine blade through the gearbox during the fault
  E_{out} is the output energy of the generator during the fault
  E_s is the energy delivered to the grid through the stator during the fault
  E_r is the energy delivered to the grid through the rotor during the fault

Because the input energy to the generator is larger than the output energies, the difference is stored in the rotating mass in the form of kinetic energy, which is expressed by

$$\Delta KE = E_m - E_{\text{out}} = \frac{1}{2} J \frac{d\omega_2}{dt} \tag{10.58}$$

where:
    ΔKE is extra kinetic energy of rotating mass (rotor of generator, gearbox, shaft, hub, and blades)
    J is the inertia constant of the rotating mass seen from the generator side of the gearbox
    ω_2 is the speed of the generator

When the ΔKE increases, the speed increases due to the positive acceleration in Equation 10.58. This could result in the blades speeding up to a level that could damage the turbine.

To restore the balance of energy, we have essentially three options, which are as follows:

1. *Option 1:* Reduce the mechanical power quickly. The problem, however, is that the mechanical power may not change fast enough. Pitching the blades takes some time as it involves rapid motion of large masses (blades, gearbox, generator, etc.). In addition, rapid actuation may cause the blades to wobble and they could hit the tower.

2. *Option 2:* Increase the electrical power output by implementing two methods.

    a. Store the extra energy somewhere outside the rotating mass.

    b. Dissipate the extra energy in external circuits.

3. *Option 3:* A combination of Option 1 and Option 2.

Storing energy during disturbances is an expensive option. It requires the use of devices that can acquire burst of energy in short time such as super capacitor and flywheel. Dissipating the extra energy during transients is a more economical option. Two techniques are discussed in this book: stator dynamic resistance (SDR) and rotor dynamic resistance (RDR).

10.5.2.1 Stator Dynamic Resistance

The SDR can be implemented by the method discussed in Chapter 8, where the power consumption of the SDR resistance is regulated through a shunt solid-state device. This method is suitable for types 1, 2, 3, and 4. For type 3 turbines, there is another method where the power consumption of the SDR resistance is not regulated by a shunt solid-state switch, but is regulated instead by controlling the rotor injected voltage. In this case, the system, shown in Figure 10.20, consists of three-phase resistance bank (R_{SDR}) connected between the generator terminals and the grid. The resistance is shunted by a bypass mechanical switch. When the system is operating normally, the bypass switch is closed and no energy is consumed by the resistance R_{SDR}. During faults, the bypass switch is opened and the injected voltage regulates the amount of energy absorbed by the resistance. Because circuit breakers in power systems operate within a few cycles, the SDR is activated for just a few milliseconds.

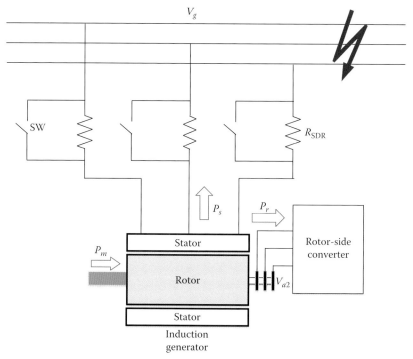

FIGURE 10.20
Stator dynamic resistance.

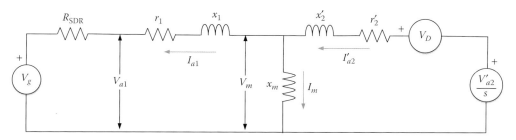

FIGURE 10.21
Equivalent circuit with stator dynamic resistance.

Consider the equivalent circuit of the DFIG in Figure 10.21. In the figure, the full value of the R_{SDR} is inserted between the generator terminals and the grid. The current in the stator circuit is

$$\overline{I}_{a1} = \overline{I}'_{a2} - \overline{I}_m = \overline{I}'_{a2} - \frac{\overline{V}_m}{jx_m} \qquad (10.59)$$

The power consumed by the resistance R_{SDR} can be computed by

$$P_{SDR} = 3\, I_{a1}^2\, R_{SDR} \qquad (10.60)$$

EXAMPLE 10.9

A DFIG system has the following parameters: $x_m = 10\ \Omega$; $r_1 = r'_2 = 0.01\ \Omega$; $x_1 = x'_2 = 0.04\ \Omega$; $N_1/N_2 = 10$.

   The generator operates at a slip of -0.2 when a grid fault occurs causing the voltage to drop to zero. The fault lasts for 100 ms. During the fault, a 1 Ω resistance is inserted between the generator and the grid. Compute the value of the injected voltage that allows the resistance to dissipate 300 kW during fault. Also, compute the energy consumed by the SDR during fault.

Solution:
The first step is to compute the stator current using Equation 10.60

$$I_{a1} = \sqrt{\frac{P_{SDR}}{3\,R_{SDR}}} = \sqrt{\frac{3\times 10^5}{3\times 1}} = 316.2\,\text{A}$$

To compute the rotor current, we need to compute the current through the magnetizing branch. Use the model in Figure 10.21 and assume $\bar{I}_{a1}$ to be the reference phasor.

$$\bar{V}_m = \bar{V}_g + \bar{I}_{a1}(R_{SDR} + r_1 + jx_1) = 0 + 316.2\times(1 + 0.01 + j0.04) = 319.36 + j12.65\,\text{V}$$

The magnetizing current is then

$$\bar{I}_m = \frac{\bar{V}_m}{jx_m} = \frac{319.36 + j12.65}{j10} = -1.265 + j31.94\,\text{A}$$

The rotor current can be computed using Equation 10.59

$$\bar{I}'_{a2} = \bar{I}_{a1} + \bar{I}_m = 316.2 - 1.265 + j31.94 = 314.935 + j31.94\,\text{A}$$

The injected voltage can be computed using the model in Figure 10.21

$$\frac{\bar{V}'_{a2}}{s} = \bar{V}_m + \bar{I}'_{a2}\left(\frac{r'_2}{s} + jx'_2\right)$$

$$\bar{V}'_{a2} = s\bar{V}_m + \bar{I}'_{a2}(r'_2 + jsx'_2)$$

$$\bar{V}'_{a2} = -0.2\times(319.36 + j12.65) + (314.935 + j31.94)(0.01 - j0.2\times 0.04) = 60.65\ \angle -175.52°\,\text{V}$$

Using the effective turns ratio, we can compute the actual injected voltage as

$$\bar{V}_{a2} = \frac{\bar{V}'_{a2}}{10} = 6.065\ \angle -175.52°\,\text{V}$$

The energy consumed by the SDR is

$$E_{SDR} = P_{SDR}\,t = 300\times 0.1 = 30\,\text{kJ}$$

If the losses (real and imaginary) in the stator and rotor windings are insignificant as compared with the ratings of the generator, we can ignore the parameters of the stator and rotor windings. In this case, we can use the approximate model shown in Figure 10.22.

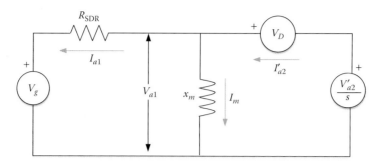

FIGURE 10.22
Approximate equivalent circuit with stator dynamic resistance.

EXAMPLE 10.10

Repeat Example 10.9 using the approximate model.

Solution:
Because $V_g = 0$, the voltage across the magnetizing branch is

$$\overline{V}_m = \overline{I}_{a1}R_{SDR} = 316.2 \times 1 = 316.2\,\text{V}$$

The magnetizing current is

$$\overline{I}_m = \frac{\overline{V}_m}{jx_m} = \frac{316.2}{j10} = -j31.62\,\text{A}$$

The rotor current can be computed using Equation 10.59

$$\overline{I}'_{a2} = \overline{I}_{a1} + \overline{I}_m = 316.2 - j31.62\,\text{A}$$

The injected voltage can be computed using the model in Figure 10.21

$$\frac{\overline{V}'_{a2}}{s} = \overline{V}_m + \overline{I}'_{a2}\frac{r'_2}{s}(1-s)$$

$$\overline{V}'_{a2} = s\overline{V}_m + \overline{I}'_{a2}\,r'_2(1-s) = -0.2 \times 316.2 + (316.2 - j31.62) \times 0.01 \times (1+0.2) = 59.45\ \angle 179.63°\,\text{V}$$

Using the effective turns ratio, we can compute the actual injected voltage

$$\overline{V}_{a2} = \frac{\overline{V}'_{a2}}{10} = 5.945\ \angle -179.63°\,\text{V}$$

Note that the detailed and approximate models produce very similar results.

10.5.2.2 Rotor Dynamic Resistance

With this technique, the dynamic resistance is inserted in the rotor circuit by opening a bypass mechanical switch, as shown in Figure 10.23. This RDR technique has two main functions: reduce the current through RSC and absorb the extra energy to balance the power equation. In the figure, note that the injected voltage V_i is not the same as V_{a2}.

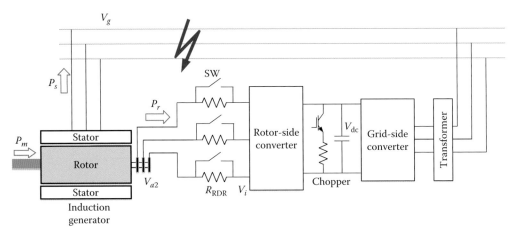

FIGURE 10.23
Rotor dynamic resistance.

If the losses (real and reactive) can be ignored, we can use one of the equivalent circuits in Figure 10.24. R'_{RDR} in the figure is the RDR resistance referred to the stator circuit. This resistance augment the existing rotor resistance and the total resistance in the rotor circuit is

$$R_{\text{tot}} = r'_2 + R'_{\text{RDR}} \tag{10.61}$$

The new developed resistance is

$$R_d = \frac{R_{\text{tot}}}{s}(1-s) \tag{10.62}$$

The developed voltage V_D is modified to account for the total resistances in the rotor circuit.

$$\bar{V}_D = -\bar{I}'_{a2}R_d = -\bar{I}'_{a2}\frac{R_{\text{tot}}}{s}(1-s) \tag{10.63}$$

The grid voltage is

$$\bar{V}_g = \bar{V}_m = \frac{\bar{V}'_i}{s} + \bar{V}_D - \bar{I}'_{a2}R'_{\text{RDR}} \tag{10.64}$$

Substituting $\bar{V}_D$ in Equation 10.63 into the above equation yields

$$\bar{V}_g = \frac{\bar{V}'_i}{s} - \bar{I}'_{a2}(R_d + R'_{\text{RDR}}) \tag{10.65}$$

The power consumed by the RDR is

$$P_{\text{RDR}} = 3(I'_{a2})^2 R'_{\text{RDR}} \tag{10.66}$$

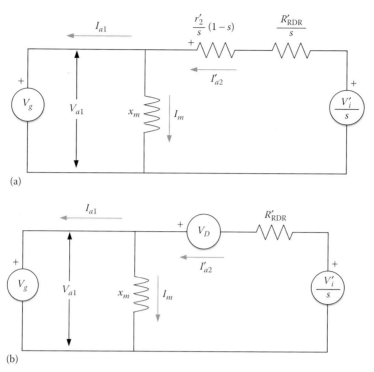

FIGURE 10.24
Models of induction generator with rotor dynamic resistance: a) Model with developed resistance and b) Model with modified developed voltage.

EXAMPLE 10.11

A DFIG system has the following parameters: $x_m = 10\ \Omega$; $r_1 = r_2' = 0.01\ \Omega$; $x_1 = x_2' = 0.04\ \Omega$; $N_1/N_2 = 10$.

The generator operates at a slip of -0.2 when a fault occurs causing the voltage at the grid side to drop to zero. The fault lasts for 100 ms. During the fault, a 10 mΩ RDR is activated. Compute the value of the injected voltage that allow the resistance to dissipate 300 kW during fault. Also, compute the energy consumed by the RDR.

Solution:
The RDR referred to the stator windings is

$$R'_{RDR} = R_{RDR} \times 10^2 = 1\ \Omega$$

The rotor current is computed from the power Equation 10.66

$$I'_{a2} = \sqrt{\frac{P_{RDR}}{3\ R'_{RDR}}} = \sqrt{\frac{300,000}{3 \times 1}} = 316.2\ \text{A}$$

The developed resistance with RDR can be computed from Equation 10.62

$$R_d = \frac{R_{tot}}{s}(1-s) = \frac{1.01}{-0.2}(1.2) = -6.06$$

The injected voltage refereed to the stator can be computed using Equation 10.65

$$\overline{V}_g = \frac{\overline{V}_i'}{s} - \overline{I}_{a2}'(R_d + R_{RDR}') = 0$$

Hence,

$$\overline{V}_i' = s\overline{I}_{a2}'(R_d + R_{RDR}') = -0.2 \times 316.2 \times (-6.06 + 1) = 310.0\,\text{V}$$

The injected voltage is in phase with the rotor current.

Using the effective turns ratio, we can compute the injected voltage in the rotor circuit

$$V_i = \frac{V_i'}{10} = 31.0\,\text{V}$$

The energy consumed by the RDR is

$$\Delta E_r = P_{RDR}\,t = 300 \times 100 = 30\,\text{kJ}$$

Exercise

1. What are the main differences between type 1, type 2, and type 3 turbines?
2. What are the advantages and disadvantages of type 3 turbines?
3. Is type 3 turbine a constant or a variable-speed system?
4. What are the factors that determine the maximum power of type 3 generators?
5. What are the factors that determine the maximum torque of type 3 generators?
6. Is the speed at the maximum torque dependent on the terminal voltage?
7. Can type 3 generator produce its own reactive power?
8. What is the slip power?
9. What are the main functions of the GSC?
10. What are the main functions of the RSC?
11. Under what condition the power flows from rotor to grid?
12. Under what condition the power flows from the grid to the rotor?
13. Can the converter provide reactive power to the grid even when the generator is not spinning?
14. What causes inrush current?
15. How is inrush current reduced?
16. What causes the generator to become unstable?
17. How is the generator controlled during grid faults?
18. How is the pullout torque adjusted?
19. What are the methods that adjust the speed of the turbine?
20. How does the crowbar system work?

21. How does the chopper work?

22. How to protect the generator from sudden sheer torques?

23. An eight-pole DFIG is spinning at 810 rpm. Its slip power is 10 kW. Ignore losses and compute the power delivered to the grid.

24. An eight-pole DFIG is spinning at 990 rpm. Its slip power is 10 kW. Ignore losses and compute the power delivered to the grid.

25. A type 3 wind turbine has a six-pole, 690 V induction generator with the following parameters: $x_1 = x_2' = 0.1\ \Omega$; $x_m = 4\ \Omega$; $r_1 = r_2' = 0.01\ \Omega$; and the stator-to-rotor effective turns ratio is 2.

    The machine is operating at a slip of −0.1. A voltage is injected into the rotor of the induction generator to increase the capacity of the system by 10% (the developed power is to increase by 10%) without changing the current or the speed of the generator. Compute the injected voltage.

26. A type 3 wind turbine has a six-pole, 690 V induction generator with the following parameters: $x_1 = x_2' = 100\ \text{m}\Omega$; $x_m = 5.0\ \Omega$; $r_1 = r_2' = 10\ \text{m}\Omega$; and stator-to-rotor turns ratio is 2.

    A voltage is injected into the rotor and the machine operates at a slip of −0.1 and is delivering 1 MW and 100 KVAr to the grid. Ignore the electrical losses and compute the injected voltage.

27. A DFIG wind turbine has a six-pole, 60 Hz three-phase, Y-connected induction generator. The terminal voltage of the generator is 690 V. The effective turns ratio of the generator is 2. The parameters of the machine are $r_1 = r_2' = 5.0\ \text{m}\Omega$; $x_1 = x_2' = 150\ \text{m}\Omega$; and $x_m = 5\ \Omega$.

    Assuming the stator voltage is the reference phasor, the injected voltage seen from the stator side is adjusted to $V_{a2}' = 5\ \angle -90°\,\text{V}$ by a PWM technique. Compute the developed power and developed torque at 1230 rpm.

28. A DFIG wind turbine has a six-pole, 60 Hz three-phase, Y-connected induction generator. The terminal voltage of the generator is 690 V. The effective turns ratio of the generator is 2. The parameters of the machine are $r_1 = r_2' = 5.0\ \text{m}\Omega$; $x_1 = x_2' = 150\ \text{m}\Omega$; and $x_m = 5\ \Omega$.

    Assuming the stator voltage is the reference phasor, the injected voltage seen from the stator side is adjusted to $V_{a2}' = 5\ \angle -90°\,\text{V}$ by a PWM technique. Compute the power delivered to the grid and the rotor injected power at 1230 rpm.

29. A 690 V, six-pole DFIG is spinning at 1300 rpm. The parameters of the machine are $r_2' = 10\ \text{m}\Omega$; $x_2' = 100\ \text{m}\Omega$; $x_m = 2\ \Omega$.

    If 2.0 V $\angle 180°$ (referred to the stator) is injected into the rotor circuit in phase with the terminal voltage, compute the developed power and the power delivered to the grid.

30. A 690 V, 10-pole DFIG system has a generator with the following parameters: $r_2' = 5\ \text{m}\Omega$; $x_1 = x_2' = 50\ \text{m}\Omega$; and $x_m = 2\ \Omega$.

    The machine operates at 760 rpm. If the injected voltage referred to the stator is $30\angle 180°\,\text{V}$, compute the reactive power delivered to the grid.

31. A 690 V, 600 rpm DFIG system has its GSC connected to the grid through a transformer whose inductive reactance is 2 Ω. Compute the output voltage of the GSC (as seen from the grid side) that would deliver 60 kVAr to the FCB.

32. A 690 V, 600 rpm DFIG system has its GSC connected to the grid through a transformer whose inductive reactance is 2 Ω. Compute the output voltage of the GSC (as seen from the grid side) that would deliver 10 kW and 60 kVAr to the FCB.

33. A DFIG system has a six-pole, 60 Hz, Y-connected induction generator. The terminal voltage of the generator is 690 V. The effective turns ratio of the generator is 5. The parameters of the machine are $r_2' = 10$ mΩ; $x_1 = x_2' = 100$ mΩ; and $x_m = 5$ Ω.

    Without injection, the machine is spinning at 1230 rpm. When the speed of the generator increases to 1260 rpm, a 5.0 V is injected in the rotor to keep the rotor current at 1.0 kA. Ignore all losses and compute the angle of the injected voltage.

34. A 10-pole DFIG is delivering 1 MW through its stator. The gearbox is delivering 1.2 MW to the generator. Ignore all losses and compute the power to the grid through the converter and the speed of the generator.

35. A dc bus of a DFIG is rated at 2 kV. The chopper circuit resistance in the dc bus is 1 Ω. The dc bus capacitor is 500 μF. Compute the duty ratio of the chopper circuit that keep the dc bus voltage to 2.1 kV and dissipate the extra energy in 1.0 ms.

36. A DFIG generator operates at a slip of −0.1 when a bolted fault (zero voltage) occurs at the terminals of the generator that lasts for 200 ms. When the fault occurs, a 1Ω SDR is inserted between the generator and the grid. The generator has the following parameters: $x_m = 5$ Ω; $r_1 = r_2' = 0.002$ Ω; $x_1 = x_2' = 0.02$ Ω; and $N_1/N_2 = 10$.

    Use a proper approximate method and compute the value of the injected voltage that allows the SDR to dissipate 3.0 MW during fault. Also, compute the energy consumed by the SDR during fault.

37. A DFIG generator operates at a slip of −0.1 when a bolted fault (zero voltage) occurs at the terminals of the generator that lasts for 200 ms. When the fault occurs, a 5 mΩ RDR is inserted in the rotor circuit. The generator has the following parameters: $x_m = 5$ Ω; $r_1 = r_2' = 0.002$ Ω; $x_1 = x_2' = 0.02$ Ω; and $N_1/N_2 = 10$.

    Use a proper approximate method and compute the value of the injected voltage that allows the RDR to limit the current to 1 kA. Also, compute the energy consumed by the RDR during fault.

11

Type 4 Wind Turbine

Most type 4 turbines are direct drive systems without gearboxes; the generator is directly coupled to the hub of the blades. Most of the generators for type 4 systems are synchronous machines with electric or permanent magnet excitation. In most designs, the magnet is mounted on the rotor, which is coupled to the hub, and the armature (where the electric power is extracted) is mounted on the stator. The magnet in this case is spinning with the blades and the armature is stationary. In newer designs, the arrangement is reversed; the magnet is mounted on the stator and is coupled to the hub (thus spinning), whereas the armature is mounted on the rotor, which is stationary.

Because the induced voltage of a moving conductor in a magnetic field depends on the rate by which the conductor is cutting the magnetic field, the frequency of the induced voltage in the armature is dependent on the speed of the blades. The relationship between the speed of the magnetic field (blades) and the frequency of the induced voltage as given in Chapter 7 is

$$f = \frac{n}{120} p \tag{11.1}$$

where:
n is the speed at which the generator is spinning (speed of the blades) in rpm
p is the number of poles of the machine
f is the frequency of the terminal voltage of the generator in Hz

To maximize the efficiency of the generator and to reduce the complexities of the converters, the generator is designed to have its output frequency near the grid frequency even at the low speed of the blades. Because the speed of the blades n is typically ranging from 6 to 60 rpm, the number of poles of the generator is large to increase f to near the grid frequency. Such a synchronous generator has typically more than 100 poles, and its cross section is very wide to fit all these poles. This is why the nacelle of type 4 system has the distinctive wide or bulged shape. One of these systems is shown in Figure 11.1. Note that the length of the nacelle is shorter than that for types 1–3 because there is no gearbox, and the width is much larger because of the large number of poles in the generator. The turbine in the figure is 6 MW offshore system with a blade length of 73.5 m. The generator is a permanent magnetic machine.

11.1 Full Converter

Because the generator rotates at the speed of the blades, the frequency at the terminals of the generator is not constant and is changing with wind speed. Therefore, the generator cannot be directly coupled to the grid. Instead, converters are used between the generator and the grid to match the frequency of the grid. One of the designs is shown in Figure 11.2. The converter is called "full converter" as it processes the full amount of power produced by the generator. It consists of three circuits: rectifier bridge, buck–boost

FIGURE 11.1
Offshore 6 MW type 4 wind turbine. (Courtesy of Alstom.)

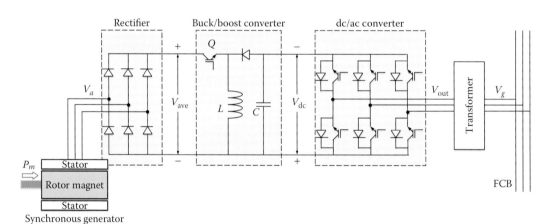

FIGURE 11.2
A type 4 system.

converter, and dc/ac converter. The rectifier bridge converts the variable frequency terminal voltage V_a into dc voltage V_{ave}. The buck–boost converter adjusts the voltage of the dc bus V_{dc}. The dc/ac converter inverts the dc voltage V_{dc} into ac voltage V_{out} with frequency that matches the grid frequency. The magnitude and frequency of V_{out} is controlled by the pulse width modulation (PWM) technique discussed in Chapter 5. The transformer is used to step up the output voltage of the converter to the grid voltage.

Based on the analysis in Chapter 5, the output of the three-phase rectifier circuit is

$$V_{ave} = \frac{3\sqrt{3}}{2\pi}\sqrt{2}\,V_a = 1.17 V_a \tag{11.2}$$

where:

V_a is the rms value of the phase voltage at the terminals of the generator

For the buck–boost converter, V_{dc} is controlled according to the following relationship:

$$V_{dc} = -V_{ave} \frac{t_{on}}{t_{off}} \qquad (11.3)$$

where:

t_{on} is the closing time of transistor Q
t_{off} is the opening time of transistor Q

As discussed in Chapter 5, the dc/ac converter with PWM has an output of

$$v_{out} = k\,V_{dc} \sin(2\pi f_s t) \qquad (11.4)$$

$$k = \frac{V_r}{V_{car}} \qquad (11.5)$$

where:

V_r is the value of the reference signal of the PWM (max or rms)
V_{car} is the value of the carrier signal (max or rms)
f_s is the frequency of the reference signal, which is set equal to the frequency of the grid
k is the gain ratio

Substituting Equations 11.2 and 11.3 into Equation 11.4 yields

$$v_{out} = -1.17\,V_a \frac{V_r}{V_{car}} \frac{t_{on}}{t_{off}} \sin(2\pi f_s t) \qquad (11.6)$$

The rms of the output voltage is

$$V_{out} = -0.827\,V_a \frac{V_r}{V_{car}} \frac{t_{on}}{t_{off}} \qquad (11.7)$$

EXAMPLE 11.1

The line-to-line voltage of a Y-connected synchronous generator is 700 V. The dc voltage of the full converter is to be adjusted to 1000 V. The switching frequency of the buck–boost converter is 5 kHz. If the output line-to-line voltage of the full converter need to be set at 600 V, compute the gain ratio of PWM circuit.

Solution:
The input to the buck–boost converter can be computed using Equation 11.2.

$$V_{ave} = 1.17\,V_a = 1.17\frac{700}{\sqrt{3}} = 472.85\ \text{V}$$

Use the switching frequency to compute the period.

$$\tau = \frac{1}{f} = \frac{1}{5000} = 0.2\,\text{ms}$$

Using Equation 11.3, compute the on time of the buck–boost transistor

$$V_{dc} = -V_{ave}\frac{t_{on}}{t_{off}} = -V_{ave}\frac{t_{on}}{\tau - t_{on}}$$

$$-1000 = -472.85\frac{t_{on}}{0.2 - t_{on}}$$

$$t_{on} = 0.136\,\text{ms}$$

$$t_{off} = \tau - t_{on} = 0.2 - 0.136 = 0.064\,\text{ms}$$

Now we can use the rms of Equation 11.4 to compute the gain ratio:

$$V_{out} = \frac{kV_{dc}}{\sqrt{2}}$$

$$V_{out} = k\frac{1000}{\sqrt{2}}$$

$$\frac{600}{\sqrt{3}} = k\frac{1000}{\sqrt{2}}$$

$$k = 0.4899$$

11.2 Power Flow

Figure 11.3 shows a type 4 system connected to the grid through a transmission line. The voltages at the output of the generator is V_a and at the output of the full converter is V_{out}, as shown in the figure. Figure 11.4 shows the one-line diagram of the system from the output voltage of the full converter as referred to the high-voltage side of the transformer (V'_{out}) to the grid bus voltage V_g.

$$V'_{out} = V_{out}\frac{N_1}{N_2} \tag{11.8}$$

where:

N_1 is the number of turns of the high-voltage winding of the transformer (grid side)
N_2 is the number of turns of the low-voltage winding of the transformer (converter side)

FIGURE 11.3
A type 4 system connected to a grid through a transmission line.

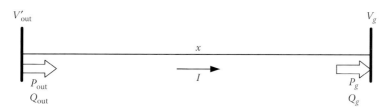

FIGURE 11.4
System in Figure 11.3 referred to high-voltage side.

The inductive reactance x in the figures is for of the transmission line x_{line} plus the transformer inductances x_{xfm}.

$$x = x_{\text{line}} + x_{\text{xfm}} \tag{11.9}$$

Because of the PWM, the magnitude and phase shift of the output voltage V_{out} are adjusted to control the flow of power to the grid. The apparent power delivered to the grid from the turbine is

$$\overline{S}_g = \overline{V}_g \, \overline{I}^* = P_g + jQ_g \tag{11.10}$$

where:
  S_g is the apparent power delivered to the grid
  P_g is the real power delivered to the grid
  Q_g is the reactive power delivered to the grid
  I is the current in the transmission line

$$\overline{I} = \frac{\overline{V}'_{\text{out}} - \overline{V}_g}{jx} \tag{11.11}$$

If we take the grid voltage V_g as a reference phasor, we get

$$\overline{I} = \frac{V'_{\text{out}}(\cos\delta + j\sin\delta) - V_g}{jx} \tag{11.12}$$

where:
  δ is the power angle (the angle between V'_{out} and V_g)

If we conjugate Equation 11.10, we get

$$\overline{S}^*_g = V^*_g \overline{I} = P_g - jQ_g \tag{11.13}$$

Substituting Equation 11.12 into 11.13 yields

$$\overline{S}^*_g = V_g \frac{V'_{\text{out}}(\cos\delta + j\sin\delta) - V_g}{jx} \tag{11.14}$$

Separating the real and imaginary components yields

$$\overline{S}^*_g = \frac{V'_{\text{out}}V_g}{x}\sin\delta - j\frac{V_g}{x}(V'_{\text{out}}\cos\delta - V_g) \tag{11.15}$$

Hence, the real power delivered to the grid is

$$P_g = \frac{V'_{out} V_g}{x} \sin \delta \qquad (11.16)$$

and the reactive power delivered to the grid is

$$Q_g = \frac{V_g}{x}\left(V'_{out} \cos \delta - V_g\right) \qquad (11.17)$$

EXAMPLE 11.2

The PWM circuit of the full converter is adjusted to produce output voltage of 750 V (line-to-line) with 10° leading angle with respect to the grid bus (infinite bus). The transformer between the full converter and the transmission line is connected in Y–Y and its turns ratio is 22. The equivalent inductive reactance of the transformer as referred to the high-voltage side is 5 Ω. The inductive reactance of the transmission line between the full converter and the infinite bus is 10 Ω. If the infinite bus voltage is 15.5 kV, compute the real and reactive powers delivered to the grid. Also, compute the reactive power produced by the full converter.

Solution:
The first step is to refer the output voltage of the full converter to the high-voltage side of the transmission line.

$$V'_{out} = V_{out} \times 22 = 750 \times 22 = 16.5\,\text{kV}$$

The total inductive reactance between the full converter and the grid bus is the sum of the transmission lines and transformer's reactive reactance.

$$x = x_{xfm} + x_{line} = 5 + 10 = 15$$

The real power for the three phases can be computed using Equation 11.16.

$$P_g = \frac{V'_{out} V_g}{x}\sin\delta = \frac{16.5 \times 15.5}{15}\sin 10 = 2.96\,\text{MW}$$

The reactive power at the grid bus can be computed using Equation 11.17

$$Q_g = \frac{V_g}{x}\left(V'_{out}\cos\delta - V_g\right) = \frac{15.5}{15}(16.5 \times \cos 10 - 15.5) = 774.3\,\text{kVAr}$$

To compute the reactive power at the terminals of the full converter. We need to compute the reactive power consumed by the transmission line and transformer.

$$\bar{I} = \frac{\bar{V}'_{out} - \bar{V}_g}{jx} = \frac{\left(16.5/\sqrt{3}\right)\angle 10° - \left(15.5/\sqrt{3}\right)}{j15} = 114\angle 14$$

The reactive power consumed by the line and transformer is

$$Q_{line} = 3I^2 x = 3 \times 114^2 \times 15 = 584.82\,\text{kVAr}$$

The output reactive power of the full converter Q_{out} is

$$Q_{out} = Q_g + Q_{line} = 774.3 + 584.82 = 1.359\,\text{MVAr}$$

11.3 Real Power Control

The equation for the real power delivered by the turbine to the grid is given in Equation 11.16. The maximum delivered power P_{max}, which is also known as the "pullout power," is the capacity of the system. It occurs at $\delta = 90°$.

$$P_{max} = \frac{V'_{out} V_g}{x} \tag{11.18}$$

Figure 11.5 shows the real power versus the power angle δ for different values of V'_{out}. As seen in the figure, the capacity of the system increases from P_{max1} to P_{max1} when the magnitude of V'_{out} increases by the PWM circuit of the dc/ac converter from V'_{out1} to V'_{out2}. Bear in mind that the increase in the capacity does not mean the delivered power increases. The only way to increase the output power of the generator is by increasing the input power from the blades.

In addition, the figure shows the case when the mechanical power from the turbine P_m is constant. If we ignore the losses, the output power of the generator P_g is constant as well. In this case, increasing V'_{out} results in a decrease in the power angle from δ_1 to δ_2, and the operating point moves from point 1 to point 2.

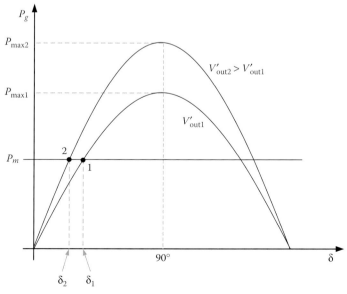

FIGURE 11.5
Real power vs. power angle for different voltages.

Because $\sin\delta \approx \delta$ for small power angles, we can approximate Equation 11.16 to

$$P_g = \frac{V_g V'_{out}}{x}\sin\delta \approx \frac{V_g V'_{out}}{x}\delta \tag{11.19}$$

EXAMPLE 11.3

A type 4 wind turbine is acquiring 2 MW from wind. The reactance of the transformer is 2 Ω and that for the transmission line is 5 Ω. The voltage at the grid bus is 13.8 kV. Compute the output voltage of the full converter as referred to the high voltage side of the transformer that operates the system at $\delta = 5°$. Also, compute the pullout power.

Solution:
Direct substitution in Equations 11.16 yields

$$P_g = \frac{V_g V'_{out}}{x}\sin\delta$$

$$V'_{out} = \frac{xP_g}{V_g\sin\delta} = \frac{(2+5)\times 2\times 10^6}{13.8\times 10^3 \times \sin(5°)} = 11.64\,\text{kV}$$

Using Equation 11.18, we can compute the pullout power

$$P_{max} = \frac{V'_{out}V_g}{x} = \frac{11.64\times 13.8}{7} = 22.95\,\text{MW}$$

11.4 Reactive Power Control

The reactive power delivered to the grid is represented by Equation 11.17 and is plotted in Figure 11.6 for different values of V'_{out}. The reactive power can be positive, zero, or negative, depending on the magnitude of voltage and the power angle.

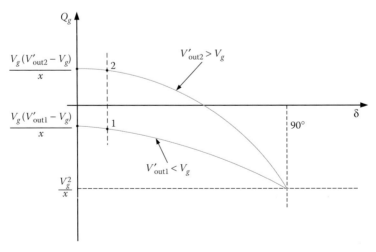

FIGURE 11.6
Reactive power vs. power angle for different voltages.

$$\text{if } V'_{\text{out}} \cos \delta < V_g, \quad Q \text{ is negative (operating point 1)}$$
$$\text{if } V'_{\text{out}} \cos \delta > V_g, \quad Q \text{ is positive (operating point 2)} \qquad (11.20)$$
$$\text{if } V'_{\text{out}} \cos \delta = V_g, \quad Q = 0$$

Positive reactive power means it flows from turbine to grid. Negative reactive power means the flow is from the grid toward the turbine.

If the system operates at small δ, $\cos \delta \approx 1$. Hence, we can approximate Equation 11.17 as

$$Q_g = \frac{V_g}{x} \left(V'_{\text{out}} - V_g \right) \qquad (11.21)$$

EXAMPLE 11.4

The power captured by the blades of a type 4 turbine is 1.3 MW. The efficiency of the turbine system is 77%. The output voltage of the converter at the high-voltage side of the transformer is 15.2 kV. The inductive reactance of the transmission line connecting the turbine to the grid plus that for the transformer is 10 Ω. The grid voltage is 15 kV. Compute the real and reactive power delivered to the grid.

Solution:

The output power of the generator is

$$P_{\text{out}} = P_{\text{blade}}\, \eta = 1.3 \times 0.77 = 1\,\text{MW}$$

If we ignore the transmission line losses, the power delivered to the grid is the output power of the generator.

$$P_g = P_{\text{out}} = 1\,\text{MW}$$

To compute the reactive power, we need to compute the power angle. This can be obtained from the power equation.

$$P_g = \frac{V_g\, V'_{\text{out}}}{x} \sin \delta$$

$$1.0 = \frac{15 \times 15.2}{10} \sin \delta$$

Hence,

$$\delta = 2.5°$$

The reactive power at the grid bus can be computed using Equation 11.17

$$Q_g = \frac{V_g}{x}(V'_{\text{out}} \cos \delta - V_g) = \frac{15}{10}\left[15.2 \cos(2.5°) - 15\right] = 278.3\,\text{kVAr}$$

Repeat the example using the small power angle approximation.

11.5 Protection

The protection of type 4 systems is essentially the same as that of type 3 systems. To protect the converters from transient overcurrent and overvoltage during system disturbances, a chopper system is used. To protect the system from becoming unstable during faults or severe disturbances, a bank of dynamic resistances is used.

11.5.1 Chopper System

During the initial few milliseconds after grid faults, the output energy E_{out} of the dc/ac converter in Figure 11.7 is reduced, while the input to the rectifier E is still high. The difference between these two energies is stored in the capacitor of the dc link, which could elevate the dc bus voltage to a level that can damage the power electronic devices of the converters. To dissipate this extra energy, a chopper is used as seen in the figure. It is designed to dissipate the extra energy in its resistance R_{ch} to maintain the voltage of the dc bus to a desired value. The duty ratio of the transistor of the chopper circuit is controlled to allow R_{ch} to consume the need amount of energy. By this method, the voltage of the dc link can be kept to a safe range.

The resistance of the chopper (R_{ch}) can be selected by a similar method to that given in Chapter 10. The first step is to compute the energy stored in the capacitor during steady-state operation.

$$E_{co} = \frac{1}{2}CV_{dco}^2 \tag{11.22}$$

where:
 E_{co} is the stored energy in the capacitor during steady-state operation
 V_{dco} is the steady-state voltage of the dc link
 C is the capacitance of the capacitor

Next, we compute the energy stored in the capacitor at the maximum allowable voltage of the dc link.

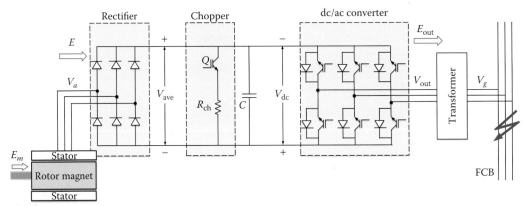

FIGURE 11.7
Chopper system.

$$E_{c-max} = \frac{1}{2}CV_{dc-max}^2 \tag{11.23}$$

where:
  E_{c-max} is the maximum stored energy in the capacitor
  V_{dc-max} is the maximum allowable voltage of the dc link

The maximum extra energy acquired by the capacitor due to the surge is

$$\Delta E_{c-max} = E_{c-max} - E_{co} = \frac{1}{2}C(V_{dc-max}^2 - V_{dco}^2) \tag{11.24}$$

The extra energy ΔE_{c-max} must be consumed by the chopper resistance. Hence,

$$E_R = P_R t = \frac{V_R^2}{R_{ch}}t = \Delta E_{c-max} \tag{11.25}$$

where:
  V_R is the voltage across the chopper resistance
  P_R is the power consumed by the chopper resistance
  E_R is the energy consumed by the chopper resistance
  t is the desired time to dissipate the extra energy

Due to the switching action of transistor Q, the voltage across the chopper resistance is

$$V_R = kV_{dc-max} \tag{11.26}$$

where:
  k is the duty ratio of the chopper's transistor

$$k = \frac{t_{on}}{\tau} \tag{11.27}$$

where:
  t_{on} is the on-time of the transistor in one period
  τ is the period of the switching cycle of the transistor

Replacing V_R in Equation 11.25 by that in Equation 11.26 yields

$$E_R = \frac{V_{dc-max}^2}{R_{ch}}\,k^2 t \tag{11.28}$$

where:
  t is the desired time to dissipate the extra energy

To keep the voltage at or near the steady-state value, the chopper resistance must consume the extra energy ΔE_{c-max}. Hence,

$$E_R = \Delta E_{c-max}$$

$$\frac{V_{\text{dc-max}}^2}{R_{\text{ch}}} \, k^2 t = \frac{1}{2} C (V_{\text{dc-max}}^2 - V_{\text{dco}}^2) \tag{11.29}$$

The duty ratio of the chopper can be computed from Equation 11.29

$$k = \sqrt{\frac{C R_{\text{ch}}}{2t}\left[1 - \left(\frac{V_{\text{dco}}}{V_{\text{dc-max}}}\right)^2\right]} \tag{11.30}$$

EXAMPLE 11.5

The rated voltage of the dc bus of a type 4 turbine is 1000 V. The dc bus capacitor is 100 μF and the chopper resistance is 2 Ω. The maximum allowed increase in dc bus voltage is 10%. Compute the duty ratio that dissipate the extra energy in 1.0 ms.

Solution:
The duty ratio of the chopper's transistor can be obtained from Equation 11.30.

$$k = \sqrt{\frac{C R_{\text{ch}}}{2t}\left[1 - \left(\frac{V_{\text{dco}}}{V_{\text{dc-max}}}\right)^2\right]} = \sqrt{\frac{10^{-4} \times 2}{10^{-3}}\left[1 - \left(\frac{1000}{1100}\right)^2\right]} = 0.186$$

11.5.2 Dynamic Resistance

When the voltage of the grid is suppressed during faults, the electrical energy output of the generator E is reduced. This reduction of the electrical energy output must be matched by a quick reduction in the mechanical energy captured by the blades. Otherwise, the turbine will accelerate and the speed of the blades could reach a destructive level. During the period from the initiation of the fault and until the blades are pitched to the correct angle, the mechanical energy going into the generator E_m in Figure 11.7 is higher than the electrical energy output of the generator E. If we ignore the losses of the windings, the energy imbalance of the generator is stored in the rotating mass of the turbine in the form of extra kinetic energy

$$\Delta KE = E_m - E = \frac{1}{2} J \frac{d\omega}{dt} \tag{11.31}$$

where:
   ΔKE is the extra kinetic energy of the rotating mass (rotor of generator, gearbox, shaft, hub, and blades)
   J is the inertia constant of the rotating mass seen from the generator side of the gearbox
   ω is the speed of the generator

When the ΔKE increases, the speed increases due to the positive acceleration in Equation 11.31. This could result in the blades speeding up to a level that could damage the turbine.

   To restore the balance of energy, we could store or dissipate the extra energy ΔKE. Dissipating the extra energy during transients can be done by the dynamic resistance shown in Figure 11.8. The system consists of a three-phase resistance bank (R_{DR}) connected to the generator terminals. The resistance is switched by a mechanical switch or solid-state

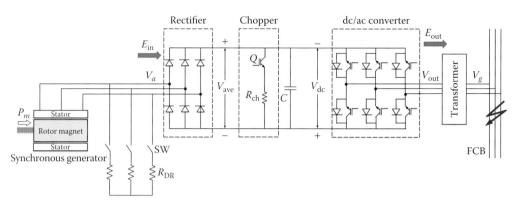

FIGURE 11.8
Dynamic resistance banks.

switch. When the system is operating normally, the switch is open and no energy is consumed by the resistance R_{DR}. During faults, the bypass switch is closed and R_{DR} absorbs the extra energy. If the bypass switch is a solid-state device, the amount of energy consumed by the resistance bank can be regulated to make $E_{in} = E_{out}$. Because most circuit breakers in the power system operate within a few cycles, R_{DR} is activated for just a few milliseconds. Keep in mind that the chopper circuit can help absorb some of the energy.

EXAMPLE 11.6

A grid fault occurs while a type 4 turbine is capturing 1 MW from wind. The dc/ac converter is disabled during fault and the extra kinetic energy is dissipated in a dynamic resistance bank. The excitation system is adjusted to keep the terminal voltage of the generator at 1.0 kV. Compute the value of the dynamic resistance.

Solution:
The dynamic resistance is required to absorb 1 MW. Hence,

$$P_{DR} = \frac{V_a^2}{R_{DR}}$$

Hence,

$$R_{DR} = \frac{V_a^2}{P_{DR}} = \frac{10^6}{10^6} = 1\Omega$$

Exercise

1. What are the main differences between type 1, type 2, type 3, and type 4 turbines?
2. What are the advantages and disadvantages of type 4 turbines?
3. Is a type 4 turbine constant or variable-speed system?

4. What are the factors that determine the maximum power of a type 4 turbine?

5. Can a type 4 turbine produce reactive power?

6. Why the buck–boost converter is sometimes used in type 4 converters?

7. Why is the converter of a type 4 system called full converter?

8. What are the main functions of the rectifier?

9. What are the main functions of the dc/ac converter?

10. How does the pullout power increase?

11. How is the reactive power controlled in type 4 turbines?

12. What causes the generator to become unstable?

13. How to protect the generator during faults?

14. A type 4 system with synchronous generator is connected to an infinite bus through a transmission line. The infinite bus voltage is 15 kV and the output voltage of the generator as referred to the high-voltage side of the transformer is 14 kV. The transmission line inductive reactance is 4 Ω, and the reactance of the transformer is 5 Ω.

   a. Compute the capacity of the system

   b. If a 2 Ω capacitor is connected in series with the transmission line, compute the new capacity of the system

15. A type 4 offshore system is connected to a 25 kV infinite bus through two parallel transmission lines. The reactance of the transformer is 2.5 Ω, and the inductive reactance of each transmission line is 2 Ω. When the output voltage of the converter is 20 kV referred to the high-voltage side, the generator delivers 10 MVA to the infinite bus at 0.8 power factor lagging. Suppose a lightning strike causes one of the transmission lines to open. Assume that the mechanical power and excitation of the generator are unchanged. Can the generator still deliver the same amount of power to the grid?

16. A type 4 system with synchronous generator is connected to an infinite bus through a transmission line. The infinite bus voltage is 15 kV. The transmission line inductive reactance is 6 Ω, and the reactance of the transformer is 4 Ω. The real power delivered to the infinite bus is 5 MW, compute the output voltage of the full converter that would deliver 1 MVAr at the infinite bus.

17. For the previous problem, the converter is connected to a transformer with turns ratio $N_1/N_2 = 10$. The voltage of the dc bus is 1400 V, the duty ratio of the chopper is 60% and its frequency is 10 kz. The rms voltage of the carrier voltage is 6 V. Compute the following:

   a. Magnitude of the reference signal

   b. Output voltage of the generator

18. The rated voltage of the dc bus of a type 4 turbine is 1.0 kV. The dc bus capacitor is 200 μF and the chopper resistance is 1 Ω. The maximum allowed increase in dc bus voltage is 20%. The duty ratio of the transistor in the chopper circuit is 0.2. How long would it take to dissipate the extra energy in the capacitor?

19. A grid fault occurs while a type 4 turbine is operating at full load. The fault lasts for 50 ms. The dc/ac converter is disabled during fault and the extra kinetic energy is dissipated in a dynamic resistance bank of 1 Ω. The excitation system is adjusted to keep the terminal voltage of the generator at 1.0 kV. Compute the power rating of the dynamic resistance and the energy consumed during fault.

12

Grid Integration

During the period from the early 1970s to the late 1980s, wind turbines (WTs) were too small to impact the operation, protection, and control of power systems. Because of wind variability and poor predictability, wind energy was not considered a significant factor in power system planning or unit commitment. Thus, their presence was unnoticed by most utilities. Curtailing wind farms and compensating owners for the loss of generation was an acceptable practice when problem occurs. For example, when wind farms caused voltage fluctuations at the generator step-up (GSU) transformers, they were curtailed. When reactive power fluctuations or reactive power demands were severe, wind farms were curtailed. When the transmission capacity at certain times could not support the variability of wind production, wind farms were curtailed. When faults occur in power system, WTs were allowed to disconnect from the grid.

From the mid-1990s, larger WTs were developed and installed in large clusters. They created wind power plants (WPPs) that are now comparable in size to conventional fossil-fuel plants. Thus, it is obvious that WPPs need to operate and be controlled like conventional power plants. Achieving this objective is a work-in-progress in most farms. The main challenges are the stochastic nature of wind and the impact of large farms on the grid. Among the technical issues that need to be addressed are the following:

- How to dispatch wind energy?
- How to use wind energy in power-system operation, for example, load following?
- How to maintain WPPs connected to the grid during system faults?
- How to support grid stability?
- How to support grid voltage?
- How to implement unit commitments?
- How to control the ramp rate of WPP?

The impacts of these problems and the possible solutions are discussed in this chapter. Keep in mind that because of the rapid advances in technology, newer solutions are introduced every year.

12.1 System Stability

The generator is an electromechanical converter. It receives mechanical power (P_m) from the turbine and delivers electrical power (P_g) to the grid, as shown in Figure 12.1. If we ignore the losses, the two energies are equal during steady-state operation.

$$P_m = P_g \tag{12.1}$$

FIGURE 12.1
Power balance.

Because the electrical power is controlled by the customers, conventional power plant continuously adjusts its mechanical power to match the electrical demand at all times. Until the powers are balanced, the surplus or deficit in power changes the kinetic energy (KE) stored in the rotating mass of the system.

$$\int_0^\tau (P_m - P_g)dt = E_m - E_g = \Delta KE \tag{12.2}$$

where:
 τ is the period when the two powers are not equal
 E_m is the mechanical energy from the turbine
 E_g is the output electrical energy of the generator
 ΔKE is the change in kinetic energy in the rotating mass (generator, turbine, etc.)

The *KE* can also be represented by

$$KE = \frac{1}{2} J\omega^2 \tag{12.3}$$

where:
 J is the moment of inertia of the rotating mass
 ω is the angular speed of the rotating mass

When the speed of the system changes from ω_1 to ω_2, the change in KE is

$$\Delta KE = \frac{1}{2} J\left(\omega_1^2 - \omega_2^2\right) \tag{12.4}$$

Examine Equations 12.2 and 12.4. When the two energies (mechanical and electrical) are equal, ΔKE is zero and the speed of the generator is constant. If the mechanical energy from the turbine is greater than the electrical energy consumed by the customers, the surplus in energy is stored in the rotating mass, thus increasing the speed of the generator. If the mechanical energy is less than the electrical energy, the deficit in energy comes from the rotating mass. Thus, the speed of the generator decreases.

12.1.1 Stability of Synchronous Generator

For synchronous generator, the speed of the generator determines the frequency of its output voltage. As shown in Chapter 7, the relationship is

$$\omega = \frac{2\pi}{60} n = 4\pi \frac{f}{p}$$

$$f = \frac{p}{4\pi} \omega \tag{12.5}$$

where:

   n is the speed of the rotating mass in rpm
   ω is the angular speed of the rotating mass in rad/s
   p is the number of poles of the machine
   f is the frequency of the generator voltage in Hz

Let us consider the case when the generator is connected to a solid grid, where the voltage and frequency are constant. Figure 12.2 shows the phasor diagram of this case, which is given in Chapter 7. In the figure, the armature resistance is ignored for large-sized generators. Each phasor in the figure has three attributes: magnitude, angle, and frequency. Because the generator is connected to a solid grid, the voltage and frequency of its terminal voltage match the frequency of the grid f_s. Because the equivalent field voltage E_f rotates at the speed of the rotor, its frequency f can change according to Equation 12.5. The power angle δ is a function of the difference between the two speeds.

$$\frac{d\delta}{dt} = \Delta\omega = \omega - \omega_s \tag{12.6}$$

If the mechanical energy is higher than the electrical energy, $f > f_s$. In this case, as revealed in Equation 12.6 and Figure 12.2, the power angle δ increases causing the vector $x_s I_a$ to increase as well. Because the synchronous reactance of the generator is fairly constant, the armature current of the generator (I_a) will increase. If the balance of energy is not restored quickly, the increase in current could reach excessive values that will result in tripping the generator by the overcurrent protection schemes. This is an unstable operation as it leads to the tripping of a generator.

If the mechanical energy is less than the electrical energy, $f < f_s$. In this case, the power angle δ decreases and E_f could lag V_a. This is a motor operation and the protection of the

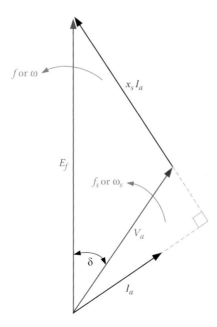

FIGURE 12.2
Phasor diagram of a synchronous generator.

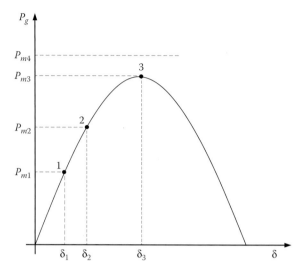

FIGURE 12.3
Power-delta curve for a synchronous generator.

generator will prevent this from happing by tripping off the generator. This is also an unstable operation.

For type 4 systems, there is a full converter between the generator and the grid, as given in Chapter 11 and shown in Figures 11.3 and 11.4. The system allows the generator to produce its energy at variable frequency while the converter keeps the frequency of its output at the grid frequency. The stability problem of this system can be explained by the power-delta curve in Figure 12.3. The curve is developed using Equation 11.16 for power in Chapter 11, which is

$$P_g = \frac{V'_{out} V_g}{x} \sin \delta \tag{12.7}$$

where:

V_g is the voltage of the grid

V'_{out} is the output voltage of the full converter referred to the high-voltage side of the transformer

x is the inductive reactance of the transformer plus the transmission line connecting the WT to the grid

P_g is the power delivered to the grid, which is the same as the mechanical power entering the turbine, as we ignored the losses

Now consider that the mechanical power of the WT is P_{m1}, and the corresponding electrical power P_{g1} is

$$P_{g1} = \frac{V'_{out} V_g}{x} \sin \delta_1 \tag{12.8}$$

The stable operation of this system is point 1, where $P_{m1} = P_{g1}$ at the power angle δ_1.

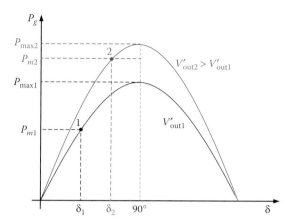

FIGURE 12.4
Increasing system capacity of a type 4 system.

If wind speed increases, the mechanical power increases to say P_{m2}. The converter then increase its power angle δ until the delivered power reaches P_{g2}. The system reaches the new operating point at 2, where $P_{m2} = P_{g2}$.

If wind power increases again to a level that matched the maximum delivered power (pullout power), the system can still theoretically reach a steady-state operating point at 3, where $P_{m3} = P_{g3} = P_{max}$. The power angle in this case is 90°.

If wind speed increases the mechanical power higher than P_{max}, the turbine cannot reach a stable operating point. This causes the machine to speed up as δ increases beyond 90°. The increase in δ results in a further reduction of the electric power delivered to the grid, thereby increasing the deficit between the mechanical power and electrical power causing the generator to speed up even faster. A stable operating point cannot be reached unless a fast control action is implemented to quickly reduce the mechanical power or increase the electrical power.

Consider the power-delta curves in Figure 12.4. Assume that the output voltage of the dc/ac converter (as referred to the high-voltage side of the transformer) is V'_{out1} and the mechanical power is P_{m1}. In this case, the system operates at point 1. Now let us assume that the mechanical power from the turbine increases to P_{m2}. In this case, $P_{m2} > P_{max1}$ and the system is unstable. However, if the output voltage of the converter is increased to V'_{out2}, which is greater than V'_{out1}, the capacity of the system increases and the system can operate at point 2.

EXAMPLE 12.1

A synchronous generator of a type 4 WT is connected to the grid through a transmission line of 0.2 pu inductive reactance. The transformer of the full converter has an inductive reactance of 0.3 pu. The voltage of the grid is 1.0 pu and the output voltage of the full converter is 1.0 pu.

1. At a mechanical power of 1.0 pu, compute the power angle that need to be adjusted by the full converter.
2. Compute the maximum power from the WT before the system becomes unstable.
3. If wind speed causes the mechanical power to increases to 2.1, will the system be stable if the converter output voltage is unchanged?

4. What is the minimum value of the output voltage of the converter that maintained the stability of the system?

5. If at $P_m = 2.1$, it is required that the converter operates at a power angle of 60°, compute the output voltage of the converter.

Solution:

1. The power can be obtained from Equation 12.8.

$$P_{g1} = \frac{V'_{out}V_g}{x}\sin\delta_1 = P_{m1}$$

The power angle is then

$$\delta_1 = \sin^{-1}\left(\frac{x\,P_{m1}}{V'_{out}V_g}\right) = \sin^{-1}\left(\frac{0.5\times1}{1\times1}\right) = 30°$$

2. The maximum power (pullout power) can be computed by setting $\delta = 90°$

$$P_{max} = \frac{V'_{out}V_g}{x} = \frac{1\times1}{0.5} = 2.0 \text{ pu}$$

3. If $P_m = 2.1$, the system is unstable as it exceeds the maximum power.

4. To stabilize the system even for $P_m = 2.1$, V'_{out} must increase, so that $P_m = P_{max} = 2.1$.

$$P_{max} = \frac{V'_{out}V_g}{x} = \frac{V'_{out}\times1}{0.5} = 2.1 \text{ pu}$$

$$V'_{out} = 1.05\,\text{pu}$$

5. To operate at a power angle of 60°, the voltage can be computed by

$$P_g = P_m = \frac{V'_{out}V_g}{x}\sin\delta$$

$$2.1 = \frac{V'_{out}\times1}{0.5}\sin 60$$

$$V'_{out} = 1.212 \text{ pu}$$

EXAMPLE 12.2

A synchronous generator of a type 4 WT is connected to the grid through a transmission line of 0.2 pu inductive reactance. The transformer of the full converter has an inductive reactance of 0.3 pu. The turbine is producing 1.0 pu of mechanical power. The voltage of the grid is 1.0 pu. The output voltage and angle of the full converter is adjusted to deliver 1.0 pu real power and 0.5 pu reactive power to the grid. Compute the voltage and power angle of the converter.

Solution:

The power from Equation 12.7 is

$$P_g = \frac{V'_{out} V_g}{x} \sin \delta = P_m$$

$$1.0 = \frac{V'_{out}}{0.5} \sin \delta$$

$$V'_{out} \sin \delta = 0.5$$

The reactive power can be computed by Equation 11.17

$$Q_g = \frac{V_g}{x}(V'_{out} \cos \delta - V_g)$$

$$0.5 = \frac{1}{0.5}(V'_{out} \cos \delta - 1)$$

$$V'_{out} \cos \delta = 0.25 + 1 = 1.25$$

Hence,

$$\frac{V'_{out} \sin \delta}{V'_{out} \cos \delta} = \frac{0.5}{1.25}$$

$$\delta = \tan^{-1} 0.4 = 21.8°$$

$$V'_{out} = \frac{1.25}{\cos \delta} = 1.346 \, pu$$

12.1.2 Stability of the Induction Generator

The stability of the induction machine can be evaluated essentially the same way as that discussed for the synchronous generator. For a type 1 system, the mechanical power from the turbine can increase up to the pullout power. Any increase in mechanical power beyond point 3 in Figure 12.5 will make the turbine unstable. This is because the developed power (which is the same as the output power if we ignore losses) is less than the acquired power from the gearbox. The turbine will speed up to high levels that could damage the WT.

For a type 2 system, the presence of the added resistance in the rotor circuit will allow us to stretch the pullout power by controlling the duty ratio of the switching circuit, as given in Chapter 9. Figure 12.6 shows the developed power of the induction generator in a type 2 system. When the duty ratio k is equal to 1, no resistance is added to the rotor circuit. When the duty ratio decreases, the pullout power increases. If the system operates at P_{m1}, the system is stable. If the mechanical power increases to P_{m2}, the system is unstable unless we reduce the duty ratio.

From the stability viewpoint, a type 2 system is more robust than a type 1 system. This is because the stable operating range of a type 2 system is wider than that of a type 1 system.

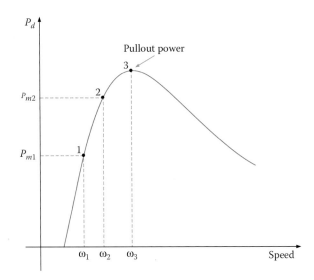

FIGURE 12.5
Mechanical power increase in a type 1 system.

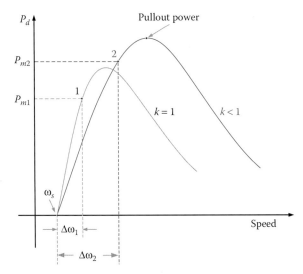

FIGURE 12.6
Mechanical power increase in a type 2 system.

EXAMPLE 12.3

A type 2 WT has a six-pole, 60 Hz three-phase, Y-connected induction generator. The terminal voltage of the generator is 690 V. The parameters of the machine are $r_2' = 10 \text{ m}\Omega$ and $x_{eq} = 0.1 \ \Omega$.

Ignore all losses and compute the stability limit of the turbine. Repeat the calculations after inserting a 0.02 Ω resistance in the rotor circuit (as referred to the stator) through a switching circuit with 50% duty ratio.

Solution:
Without the added rotor resistance, we compute the slip at pullout power, as given in Equations 9.23 and 9.24.

$$m = \frac{r_2'}{x_{eq}} = 0.1$$

$$s' = -m\left[m + \sqrt{m^2 + 1}\right] = -0.1 \times \left[0.1 + \sqrt{0.1^2 + 1}\right] = -0.1105$$

The stability limit is the pullout power (or maximum power) in Equation 9.25

$$P_{d-max} = -3\frac{(1-s')}{s'}\frac{V_{a1}^2 r_2'}{\left(r_2'/s'\right)^2 + (x_{eq})^2} = 3\frac{(1+0.1105)}{0.1105}\frac{(690^2/3)0.01}{(0.01/-0.1105)^2 + (0.1)^2} = 2.53\,\text{MW}$$

When the resistance is added, we need to compute r_{add} based on the duty ratio in Equation 9.1.

$$r_{add} = r(1-k) = 0.02 \times (1-0.5) = 0.01\,\Omega$$

Now we can compute the slip at the pullout power.

$$m = \frac{r_2' + r_{add}'}{x_{eq}} = \frac{0.01 + 0.01}{0.1} = 0.2$$

$$s' = -m\left[m + \sqrt{m^2 + 1}\right] = -0.2 \times \left[0.2 + \sqrt{0.2^2 + 1}\right] = -0.244$$

The new stability limit is

$$P_{d-max} = -3\frac{(1-s')}{s'}\frac{V_{a1}^2(r_2' + r_{add}')}{\left[(r_2' + r_{add}')/s'\right]^2 + (x_{eq})^2} = 3\frac{(1+0.244)}{0.244}\frac{(690^2/3)0.02}{(0.02/-0.244)^2 + (0.1)^2} = 2.904\,\text{MW}$$

A type 3 system can achieve better stability range than types 1 and 2 systems. The range is adjusted by the injected voltage in the rotor circuit, as shown in Figure 12.7. If the mechanical power increases from P_{m1} to P_{m2}, the system is unstable unless we change the injected voltage to operate the system at point 2.

12.1.3 Systemwide Stability

The normal operation of a power system is when the power generated from all sources are consumed by all loads plus system losses at all times. This delicate balance between generation and demand can be described by the conceptual graph in Figure 12.8. The lack of considerable energy storage in power grids causes the surplus in generation to speed up the turbines of the power system. In addition, the deficit in generation slows down the turbines. If the imbalance is not compensated quickly, the change in speed (frequency) could trip some turbines causing a wider imbalance that could eventually lead to blackouts. The system in this case is judged to be unstable. Power system stability is defined as the ability

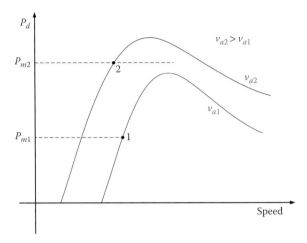

FIGURE 12.7
Mechanical power increase in a type 3 system.

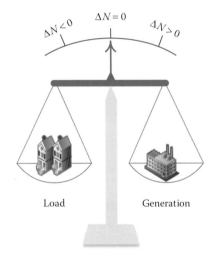

FIGURE 12.8
Power balance.

of the power system to achieve the steady-state operating condition after a disturbance. This means that all oscillations (frequency, voltage, power, etc.) must be damped out, as shown in Figure 12.9. In case of unstable system, the oscillations are not damped out and the system behaves, as shown in Figure 12.10 or Figure 12.11.

With conventional turbines, the balance of energy is restored mainly by controlling the mechanical power of the turbines. If this action is not enough to restore the balance of powers, the alternative is to shed some loads creating controlled outages. However, if the system becomes stable after a disturbance, it could still be insecure. Insecure system is the one that has some of its major equipment (generator, transformer, and transmission line) operating beyond their design limits (rated values): bus voltages could be lower or higher than the set limit, transmission lines could carry excessive currents, or transformers could operate beyond their thermal limits. In these cases, the system could face additional

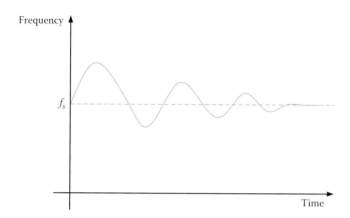

FIGURE 12.9
Stable system.

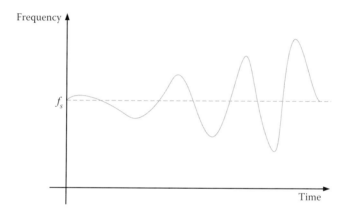

FIGURE 12.10
Unstable system (lack of system damping torque).

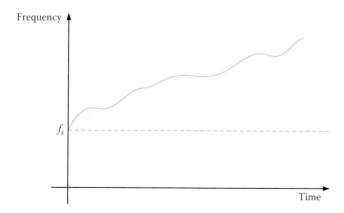

FIGURE 12.11
Unstable system (lack of synchronizing torque).

outages. The power system is only secure if the system is stable after a disturbance and its key components are operating within their design limits.

With the high penetration of wind, the power system operator (PSO) has little control on the mechanical power of the WTs, mainly because wind is intermittent and stochastic. If a temporary disturbance occurs that require an increase in wind energy output, it may not be achieved unless favorable wind conditions exist.

Because WPP are becoming larger in size, their impact on the grid is substantial. Therefore, several governmental regulatory agencies are requiring the WPP to operate in similar way as conventional power plants. Among these organizations is the Federal Energy Regulatory Commission (FERC) in the United States, which has established dynamic performance requirements for WTs to help support grid security and stability. This would mean that WPP develops a system by which production can be controlled based on commands from the PSO. Different active power regulation functions are required in wind farms, among them are the ones discussed subsequently and shown in Figure 12.12.

Delta control: With delta control, the WPP produces less power than possible at any wind condition by a margin Δ. This way, when the PSO needs to increase system generation, the WPP can participate by ramping up its production by up to Δ. The drawback of this technique is the spilling of wind power at all times. Alternatively, WPP can install fast acting generation units, such as gas turbines, and use them to ramp up the production of the plant when needed.

Maximum production constraint: At low power demands (nights, for instance), WPP limits their production to a maximum limit P_{max}. This way, surplus in power generation in the grid is avoided. This technique is preferred by some WPP producers over the direct curtailment of the plant production.

Balance control: WPP production is regulated by the operator according to system needs. This scenario would require WPP to be controlled by the grid control centers.

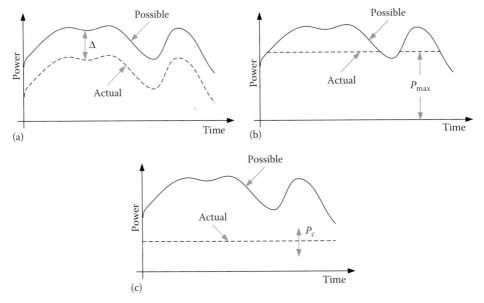

FIGURE 12.12
Power regulation functions: (a) delta control, (b) maximum production constraint, and (c) balance control.

Besides the above functions, several utilities have requirements on how fast or slow is the ramping of power produced by WPPs. Some of the regulations are as follows:

Rate of ramping up—WPP are required to comply with ramp-up rates set by PSOs and are based on grid conditions. For example, after clearing grid faults, ramping up the production has to be fast to balance the power, thus prevent forced outages to customers. However, under normal operating conditions, utilities demand slow ramping when wind speed increases to allow the PSOs to accommodate the extra energy production in their systems.

Rate of ramping down—During normal operation, ramping down is required to be slow to allow the PSOs to compensate for the deficit in energy production in their systems. However, during faults, WPP operators ramp down the turbines as fast as possible to prevent damaging WTs. Keep in mind that excessively fast ramp rates may cause the blades to wobble and hit the tower.

12.2 Fault Ride-Through, Low-Voltage Ride-Through

Power system is continuously subjected to faults. The majority of the faults are temporary and the rest are permanent. Temporary faults are caused by events such as trees brushing power lines, birds and animal intrusions, and insulations flashover. Temporary faults are cleared by themselves often after a few milliseconds. Permanent faults are more serious and are often caused by insulators failure, downed conductors, earthquake, falling trees on power lines, internal shorts of equipment, and so on. Removing the source of permanent faults takes long time, as they require workers to perform maintenance on the damaged parts.

Utilities operate with the assumption that most faults are temporary and can be cleared quickly by themselves. Thus, the turbines in the power system stay connected to the grid to generate electricity immediately after the temporary fault is cleared. This way, the balance of power can be achieved and customers' service is restored quickly.

Historically, WTs were allowed to disconnect from the grid when faults occurred to protect them from being damaged due to over speeding. However, with the larger penetration of wind energy, tripping large number of wind generators can result in a large deficit in generation after the fault is cleared, thus contributing to the instability of the power system. Because of this problem, WPPs are now required to remain connected to the grid during temporary faults or until the protection devices clear permanent faults. This is called "fault ride-through" (FRT). In a more general description, it is called "low-voltage ride-through" (LVRT).

12.2.1 Impact of Fault on WTs

The impacts of grid faults on WTs employing induction generators differ from these employing synchronous generators. Because types 1 and 2 induction generators receive their excitation from the grid, bolted faults (near zero voltage) make these generators lose their electromechanical conversion capability. Thus, cannot stay connected to the grid. For type 3 generators, the doubly fed induction generator can be designed to provide some excitation during a short period due to the presence of the capacitor in the dc bus, which will allow the generator to remain connected to the grid. For a type 4 system, the excitation is not lost and the generator can stay connected to the grid. To understand how WPP can achieve LVRT requirements, let us discuss the impact of faults on WTs.

12.2.1.1 Current

Under normal operating conditions, the mechanical power from the turbine is converted into electrical power that is eventually delivered to the grid at the point of interconnection (POI), as shown in Figure 12.13. If we ignore the electrical losses, all these powers are equal.

For the system in Figure 12.13, the simplified equivalent circuit, which is also given in Chapter 6, is shown in Figure 12.14. In the figure, x_{tl} represents the inductive reactance of the transformer plus the trunk line. V_{poi} is the voltage at the POI. In this equivalent circuit, all losses are ignored.

Now let us assume a bolted fault (grounded three-phase fault with zero impedance) occurs at the POI, where V_{poi} collapses to zero. After the initial transients subside, the fault current in the rotor is

$$\bar{I}'_{a2} = \frac{\bar{V}_D + \left(\bar{V}'_{a2}/s\right)}{\bar{x}} \tag{12.9}$$

where:

$$x = x'_2 + x_{eq}$$
$$x_{eq} = \frac{x_m(x_1 + x_{tl})}{x_m + x_1 + x_{tl}} \tag{12.10}$$

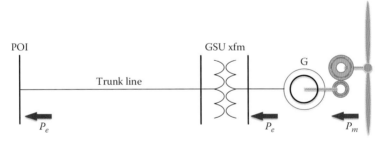

FIGURE 12.13
Real power production.

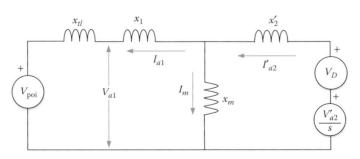

FIGURE 12.14
Simplified model of the system in Figure 12.13.

The developed voltage is,

$$\bar{V}_D = -\bar{I}'_{a2} \frac{r'_2}{s}(1-s) \tag{12.11}$$

Hence, the steady-state fault current in the rotor is

$$\bar{I}'_{a2} = \frac{(\bar{V}'_{a2}/s)}{(r'_2/s)(1-s) + jx} \tag{12.12}$$

and the stator current is

$$\bar{I}_{a1} = \bar{I}'_{a2} \frac{x_m}{x_m + x_1 + x_{tl}} \tag{12.13}$$

As seen in Equations 12.12 and 12.13, the magnitude of the fault current depends on the injected voltage. For type 1 or type 2 systems, there is no injection and the fault current is zero. For a type 3 system, the injected voltage V_{a2} produces fault current with adjustable magnitude. This feature allows the turbine to remain electrically connected to the grid during temporary faults to provide some reactive power support, as seen in Section 12.2.1.2, thus enhancing the stability of the power system.

EXAMPLE 12.4

A 690 V, six-pole, Y-connected, type 3 generator has the following parameters: $r'_2 = 10$ mΩ; $x_1 = x'_2 = 100$ mΩ; $x_m = 5$ Ω

The inductive reactance of the trunk line plus transformer as referred to the low voltage side is 1.0 Ω, and the phase value of the injected voltage referred to the stator is 10 V. The machine was spinning at 1260 rpm when a bolted fault occurred at the POI. Ignore all electrical losses and compute the steady-state current feeding the fault.

Solution:
Compute the various inductive reactances

$$x_{eq} = \frac{x_m(x_1 + x_{tl})}{x_m + x_1 + x_{tl}} = \frac{5 \times 1.1}{5 + 0.1 + 1} \approx 0.9 \ \Omega$$

$$x = x'_2 + x_{eq} = 1.0 \ \Omega$$

Compute the slip

$$s = \frac{1250 - 1200}{1200} = -0.05$$

The current can be obtained from Equation 12.12

$$\bar{I}'_{a2} = \frac{(\bar{V}'_{a2}/s)}{(r'_2/s)(1-s)+jx} = \frac{-(10/0.05)}{(0.01/-0.05)(1.05)+j1.0} = 195.73 \angle 78.14° \text{ A}$$

$$\bar{I}_{a1} = \bar{I}'_{a2} \frac{x_m}{x_m+x_1+x_{tl}} = 195.73 \angle 78.14° \times \frac{5}{6.1} = 160.43 \angle 78.14° \text{ A}$$

12.2.1.2 Reactive Power

After the temporary fault is cleared, a large amount of reactive power is rushed into the induction generator due to its inductive reactances. For types 1 and 2 turbines, the surge of the reactive powers comes from the grid and can depress the voltage in the system, which can prolong or hinder the recovery of the power system after the fault is cleared. For types 3 and 4 systems, the generators can compensate for the reactive power consumed by the various inductive reactances in the system during and after faults. This is a great voltage support feature that makes types 3 and 4 systems more friendly to grid integration.

Let us assume a bolted fault occurs at the POI, as given in Figure 12.15 (also see Figure 12.16). In this case, the reactive powers consumed by the various inductive reactance components is

$$Q = Q_{tl} + Q_1 + Q_m + Q_2 \tag{12.14}$$

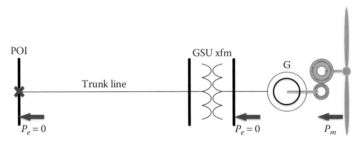

FIGURE 12.15
Real power collapse during bolted fault at POI.

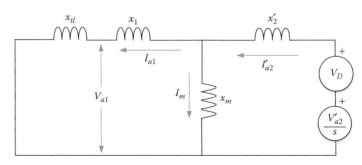

FIGURE 12.16
Simplified model of the system in Figure 12.15.

where:
   Q is the total reactive power consumed by all inductive reactances during fault
   Q_{tl} is the reactive power consumed by the transmission line
   Q_1 and Q_2 are the reactive power consumed by the stator and rotor windings, respectively
   Q_m is the reactive power consumed by the magnetizing branch

$$Q = I_{a1}^2 x_{tl} + I_{a1}^2 x_1 + I_m^2 x_m + I_{a2}'^2 x_2' \tag{12.15}$$

$$\bar{I}_{a2}' = \frac{(\bar{V}_{a2}'/s)}{jx}$$

$$\bar{I}_{a2}' = \frac{(\bar{V}_{a2}'/s)}{(r_2'/s)(1-s) + jx} \tag{12.16}$$

$$\bar{I}_m = \bar{I}_{a2}' \frac{x_1 + x_{tl}}{x_m + x_1 + x_{tl}}$$

The total three-phase consumption of reactive power during fault is

$$Q = 3I_{a1}^2(x_{tl} + x_1) + 3I_m^2 x_m + 3I_{a2}'^2 x_2' \tag{12.17}$$

As seen in Equations 12.12 and 12.13, the current during fault is controlled by the injected voltage. Hence, the magnitude of the reactive power consumed by the various inductive reactances is also controlled by the injected voltage.

EXAMPLE 12.5

A six-pole type 3 WT generator is connected to the grid through a trunk line. The parameters of the generator are $r_2' = 10$ mΩ; $x_1 = x_2' = 100$ mΩ; $x_m = 5$ Ω.
   The inductive reactance of the truck line plus transformer as referred to the low-voltage side of the GSU transformer x_{tl} is 2 Ω. During normal operation, the generator is spinning at 1260 rpm, and reactive power consumed by all inductances is 1.0 MVAr.
   A bolted fault occurs at the POI. Compute the injected voltage that provides 1.0 MVAr to all inductive reactances.

Solution:
Find the current relationships using Equation 12.16

$$\bar{I}_{a1} = \bar{I}_{a2}' \frac{x_m}{x_m + x_1 + x_{tl}} = \frac{5}{5 + 0.1 + 2} \bar{I}_{a2}' = 0.704 \bar{I}_{a2}'$$

$$\bar{I}_m = \bar{I}_{a2}' \frac{x_1 + x_{tl}}{x_m + x_1 + x_{tl}} = 0.296 \bar{I}_{a2}'$$

Substituting the currents into Equation 12.17 yields the rotor current

$$Q = 3I_{a2}'^2 \times 0.704^2 \times 2.1 + 3I_{a2}'^2 \times 0.296^2 \times 5 + 3I_{a2}'^2 \times 0.1 = 10^6 \text{ VAr}$$

$$4.7366 I_{a2}'^2 = 10^6 \text{ VAr}$$

$$\overline{I}_{a2}' = 459.5 \text{ A}$$

Compute the slip when the fault occurs

$$s = \frac{1250 - 1200}{1200} = -0.05$$

Compute x of Equation 12.10

$$x = x_2' + \frac{x_m(x_1 + x_{tl})}{x_m + x_1 + x_{tl}} = 1.58$$

The injected voltage can be computed from Equation 12.16

$$\overline{V}_{a2}' = s\overline{I}_{a2}' \left(\frac{r_2'}{s}(1-s) + jx \right) = -0.05 \times 459.5 \times \left(\frac{0.01}{-0.05}(1.05) + j1.58 \right) = 36.62 \angle -82.43° \text{ V}$$

Note that the current leads the injected voltage as the rotor side converter (RSC) is delivering reactive power.

12.2.1.3 Mechanical Stress

Faults on power systems cause mechanical stress to the turbines. The mechanical stress could damage the drive train of the turbine and even the structure of the turbine itself due to excessive over speeding and excessive torsional torques. The severity of the stress depends on several factors such as the location of the fault with respect to the turbine, the voltage reduction during fault, and the power captured by the turbine from wind.

Consider the drive train of the WT shown in Figure 12.17. The system consists of rotating blades, low-speed shaft, gear, high-speed shaft, and the generator. The lift torque from the blades acts on one end of the low-speed shaft, while the developed torque from the generator side acts on the other end of the high-speed shaft. The difference between the two torques is the torsional torque on the shafts. In the steady state, all torques are in equilibrium and the torsional torque is zero. Because the high-speed shaft is often short (thus stiff), we can assume that the torsional torque is mainly on the low-speed shaft. Excessive torsional torque on drive shafts is the main cause of failure in rotating mechanical systems.

During grid faults, the output power of the generator is reduced considerably, roughly by the square of its terminal voltage. Thus, the developed torque is significantly reduced, and thereby the torsional torque is considerably increased. The various torques and speed of the drive train start to oscillate. Unless it is damped out and controlled, the oscillation could cause the shaft to break.

To understand the dynamics of the drive train, let us start with the kinetic energy (*KE*) of any rotating mass

$$KE = \frac{1}{2} J \omega^2 \tag{12.18}$$

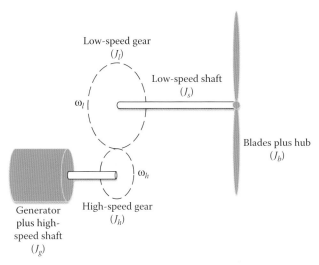

FIGURE 12.17
Drive train of wind turbine.

where:
  J is the moment of inertia of the rotating mass
  ω is the angular speed of the shaft

For the system in Figure 12.17, the total *KE* is the sum of the *KE* of the main components

$$KE_{\text{total}} = KE_{\text{blade}} + KE_{\text{shaft}} + KE_{\text{gear}} + KE_g \qquad (12.19)$$

where:
  KE_{blade} is the KE of the blades and hub
  KE_{shaft} is the KE of the low-speed shaft
  KE_{gear} is the KE of the gear system
  KE_g is the KE of the generator and high-speed shaft
  KE_{total} is the total KE of the system

The *KE* of the entire system can be represented by an equivalent mass rotating at the speed of the shaft

$$KE_{\text{total}} = \frac{1}{2} J_{\text{eq}} \omega_l^2 \qquad (12.20)$$

where:
  J_{eq} is the equivalent inertia of the system seen from the low-speed shaft
  ω_l is the speed of the blades

Hence,

$$KE_{\text{total}} = \frac{1}{2} J_{\text{eq}} \omega_l^2 = \frac{1}{2} J_g \omega_h^2 + \frac{1}{2} J_h \omega_h^2 + \frac{1}{2} J_l \omega_l^2 + \frac{1}{2} J_s \omega_l^2 + \frac{1}{2} J_b \omega_l^2 \qquad (12.21)$$

where:

ω_h is the speed of the generator (high speed shaft)
J_b is the inertia of the blades and hub
J_s is the inertia of the low-speed shaft
J_l is the inertia of the wheel of the low-speed gear
J_h is the inertia of the wheel of the high-speed gear
J_g is the inertia of the generator and high-speed shaft

Hence, the equivalent inertia is

$$J_{eq} = J_g \left(\frac{\omega_h}{\omega_l} \right)^2 + J_h \left(\frac{\omega_h}{\omega_l} \right)^2 + J_l + J_s + J_b \tag{12.22}$$

If we ignore the mechanical losses, the developed power can be computed at either the low-speed or high-speed shafts.

$$P_d = T_{eh}\omega_h = T_{el}\omega_l \tag{12.23}$$

where:

P_d is the developed power of the generator
T_{eh} is the electric torque computed at the high-speed shaft (developed torque of the generator)
T_{el} is the electrical torque computed at the low-speed shaft

Hence,

$$T_{el} = T_{eh} \frac{\omega_h}{\omega_l} \tag{12.24}$$

The dynamics of the rotating drive train can be represented by the second order spring equation

$$T_{blade} - T_{el} = \Delta T = J_{eq} \frac{d^2\theta}{dt^2} + D\frac{d\theta}{dt} + K\theta \tag{12.25}$$

where:

T_{blade} is the mechanical torque produced by the blades
θ is the angle of the twist (torsional torque angle). It is the difference between the angles at both ends of a shaft
D is the damping coefficient (due to system friction, windage, etc.)
K is torsion coefficient
ΔT is the torsional torque

The angle of the twist is a function of the difference in speeds at both ends of the shaft ($\Delta\omega$)

$$\frac{d\theta}{dt} = \Delta\omega \tag{12.26}$$

Equation 12.25 can be rewritten in the general second order form

$$\frac{\Delta T}{J_{eq}} = \frac{d^2\theta}{dt^2} + 2\xi\omega_n\frac{d\theta}{dt} + \omega_n^2\theta \tag{12.27}$$

where:

ω_n is the speed of natural oscillations

ω_n is the speed of natural oscillations

ξ is the damping coefficient

$$\omega_n = \sqrt{\frac{K}{J_{eq}}}$$

$$\xi = \frac{D}{2}\sqrt{\frac{1}{K J_{eq}}}$$

(12.28)

Based on the values of the damping coefficient and natural oscillations, the dynamic performance of the system can be determined. A good control system is the one that adjusts the pitch angle of the blade in such a way as to enhance the system damping and reduces the stress on the system.

Figure 12.18 shows several of these dynamic oscillations. In Figure 12.18a, the system has no damping and the natural frequency of oscillation is high. This occurs when the inertia of the system is small and system damping is small. In Figure 12.18b, the damping coefficient is larger and the natural oscillation is lower. This occurs when the torsion coefficient is small or the damping coefficient is large. In Figure 12.18c, the torsional torque could exceed the design limit of the system, which could cause the shaft to break.

Keep in mind that the magnitude of the first swing depends on the magnitude of the disturbance. The worst scenario is when a bolted fault occurs while the turbine is producing maximum torque. For type 1 and 2 turbines, the electric torque collapses to zero, hence,

$$\Delta T = T_{blade-max} - T_e \approx T_{blade-max}$$

$$\frac{\Delta T}{J_{eq}} = \frac{T_{blade-max}}{J_{eq}} = \frac{d^2\theta}{dt^2} + 2\xi\omega_n\frac{d\theta}{dt} + \omega_n^2\theta$$

(12.29)

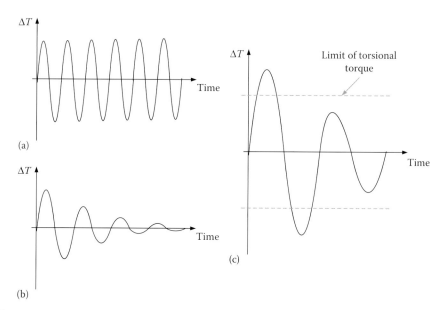

FIGURE 12.18

Mechanical dynamics: (a) small damping; (b) higher damping; and (c) exceeding limit on torsional torque.

12.2.2 LVRT Requirements

If large WPPs are allowed to disconnect from the grid during low-voltage events, they may create a large deficit in energy that could lead to blackouts. Therefore, large WTs are required to remain connected to the grid if the voltage is above a time-dependent profile called low-voltage ride-through (LVRT) profile, or fault ride-through (FRT) profile.

A generic LVRT requirement (profile) is shown in Figure 12.19. If the actual per unit voltage during fault at the GSU transformer (or at any other agreed upon location) is above this curve, WTs must remain connected to the grid. V_{min} in the figure is the minimum voltage during fault. V_r is the recovery voltage after the fault is cleared, t_c is the clearing time of fault. t_{end} is the end time of the low-voltage event. All these parameters are determined by reliability codes, fault-clearing capabilities, and the interconnection requirements study made by utilities to accommodate new wind generation within their service areas.

In the United States, Federal Energy Regulatory Commission (FERC) established interconnection requirements for WPP. In its Order No. 661-A issued in December 12, 2005, FERC required all generating units, including renewable, to remain connected to the grid after a fault on any or all phases for the time it takes to clear the fault (about 9 cycles). The voltage during fault could be as low as zero volts at the high voltage (HV) side of the GSU transformer.

Figure 12.20 shows samples of LVRT requirements. In the figure, the fault occurs at $t = 0$. These requirements often change when newer technologies are available. For example, FERC in the United States initially set V_{min} to 15%, then reduces it to zero for newer installations.

Keep in mind that the induction generators for types 1 and 2 turbines relay on the grid to produce their magnetic fields. Hence, they require a minimum voltage at their terminals to remain connected to the grid. Thus, they cannot ride through bolted faults. However, because types 3 and 4 systems can produce their magnetic fields from auxiliary sources, they can remain connected to the grid even during bolted faults.

In Figure 12.21, the LVRT requirement and the actual system voltage at the POI during fault are shown. In this case, the system voltage due to the disturbance is higher than the LVRT requirement curve at all times. Thus, the WT is not allowed to disconnect from the grid. However, for the case in Figure 12.22, the actual voltage during the disturbance is dipped below the LVRT curve. Thus, the turbines are allowed to trip at t_{trip}.

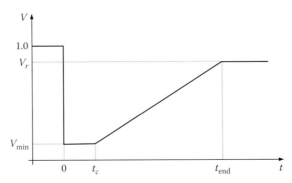

FIGURE 12.19
A generic LVRT requirement.

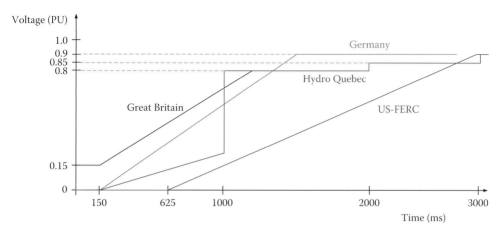

FIGURE 12.20
Samples of an LVRT requirement.

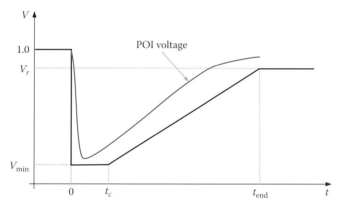

FIGURE 12.21
Example of a condition at which wind turbine is not allowed to trip.

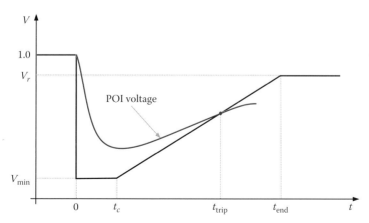

FIGURE 12.22
Example of a condition at which wind turbine is allowed to trip.

The trunk lines of most large wind farms are connected to the high-voltage transmission lines. However, some wind farms are increasingly connected to distribution systems. Also, it is not uncommon for the trunk lines of wind farms to share the same towers with distribution lines. In these cases, wind farms are regularly subjected to distribution system faults, which have different fault-clearing procedures than these for transmission line faults. For transmission lines, as stated in IEEE Standard 37.102, fault-clearing methods and reclosing have four basic features:

1. Reclosing circuit breakers are delayed by at least 10 seconds.
2. Reclosing circuit breakers starts from the remote end of a faulted line to reduce the stress on the turbines.
3. Reclosing is done often once.
4. Further automatic reclosing is block at the generating station if the first attempt fails.

For distribution networks, however, the philosophy is different: automatic reclosing is after a much shorter time, and it can be repeated a number of times within a few minutes. This is because of the assumption that most faults are temporary and can be cleared by themselves. If the fault is persistent, repeated automatic reclosing has the same effect as multiple faults. Therefore, LVRT requirements in some regions require ride-through multiple faults capabilities. To date, there is no consensus on how to achieve this capability. The following three capabilities are proposed:

Single-fault capability: A WT is not required to ride though multiple faults within short time. This argument is resisted by utilities as it means the WPP is disconnected for most of the distribution network faults

Multiple fault, single LVRT capability: A WT is required to be connected if the multiple faults occur within one LVRT curve, as shown in Figure 12.23. The turbine is allowed to trip when the voltage at any instant falls below the LVRT curve, at t_{trip} in Figure 12.24.

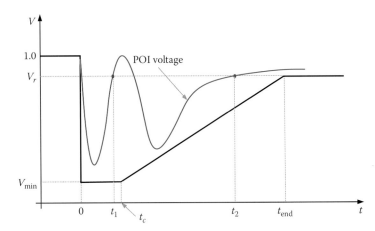

FIGURE 12.23
Example of a multiple fault at which wind turbine is not allowed to trip.

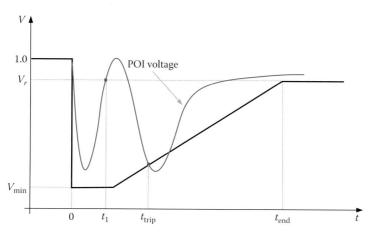

FIGURE 12.24
Example of a multiple fault at which wind turbine is allowed to trip under single LVRT rule.

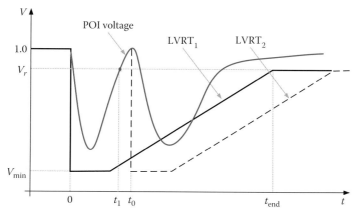

FIGURE 12.25
Example of a multiple fault at which wind turbine is not allowed to trip under multiple LVRT rule.

Multiple fault, multiple LVRT: Each of the multiple faults is treated as a different event, as shown in Figure 12.25. If due to the initial fault, the voltage recovers to V_r at t_1, the ride-through event is completed. When the second fault occurs, it is treated as a new event with new LVRT curve. One major technical challenge to this method is how fast we can accurately compute the rms voltage if reclosing occurs within a cycle.

12.2.3 LVRT Compliance Techniques

In order for the WPP to comply with the LVRT requirements, various controls strategies can be implemented. Some of them control the speed of the turbine, and others use external equipment in addition to the speed control. The use of the external equipment requires some excitation in the generator. Hence, it cannot work for types 1 and 2 systems during bolted faults. The control strategy could be decentralized for individual turbines or centralized for a cluster of turbines. Some of these methods are discussed subsequently.

12.2.3.1 Ramping Control

WTs ramp continuously by adjusting the pitch angle of their blades due to wind fluctuates or control actions. In normal operation, WT ramps are limited to allow conventional generators to adjust their output and maintain the energy balance of the grid. However, in emergency operation, WT ramps are required to be large enough to maintain system stability. During faults, WTs must adjust their blades quickly to reduce the mechanical energy captured from wind. Then, right after fault is cleared, the turbine must ramp up its production rapidly to maintain system stability and security. Some of the international standards require WTs to ramp up to as much as 90% of its prefault production within a few seconds. The maximum ramping speed by pitch control depends on the length of the blade and their material. This is because excessively fast pitch control can cause flexible long blades to wobble and hit the tower.

  If the ramping rate of the WT is less than desired during emergency, other methods can be implemented at the WPP facilities. For example, a fast-acting gas turbine generator or energy storage can be used for ramping up the WPP. For ramping down, the excessive energy can be consumed locally and mechanical brakes can be used. Some of these methods are discussed subsequently.

12.2.3.2 Dynamic Braking

The extra kinetic energy stored inside the drive train during faults could speed up the blades to a level that could damage the turbine unless the extra energy is dissipated quickly. This can be accomplished by converting the extra kinetic energy into electrical that can be either consumed by a braking resistance, or stored in batteries, flywheels, and so on.

  The concept of dynamic braking is shown in Figure 12.26. The system consists of a braking resistance (R_b) connected between the generator and the POI. The resistance is bypassed by the anti-parallel solid-state switch that is closed during normal operation. When faults occur, the bypass switch is opened and regulated, so that the braking resistance absorbs enough energy to prevent the turbine from over speeding. During braking, the turbine ramps down by pitch control until a balance of energy is achieved. This technique requires excitation to the generator, which is possible for types 3 and 4 systems.

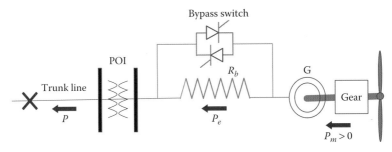

FIGURE 12.26
Series braking resistance circuit.

EXAMPLE 12.6

A 690 V, 2 MW WT operating at full power output when bolted fault occurred at the POI. Size the series braking resistance.

Solution:
To size the resistance, we compute the maximum energy consumption of the braking resistance. Assume that it operates for 2 seconds during bolted fault. Hence, the maximum energy to be absorbed is

$$E = Pt = 2 \times 2 = 4\,\text{MJ}$$

For bolted fault, the voltage drop across the resistance can be assumed to be the full voltage at the terminals of the generator. Hence, the ohmic value of the braking resistance is

$$R_b = \frac{V^2}{P} = \frac{690^2}{2 \times 10^6} = 0.24\,\Omega$$

The series braking resistance described in Figure 12.26 requires the solid-state bypass switch to be closed at all times except during faults. The drawback of this method is that the losses due to the continuous operation of the switch reduce the overall efficiency of the turbine. An alternative is to use the shunt dynamic braking method shown in Figure 12.27. In this system, the solid-state switch is opened during normal operation and is closed only during fault. The losses in this system are much smaller than those for series braking method. However, the shunt resistance is not effective for bolted faults. This is because the power absorption by the braking resistance is a function of the square of the terminal voltage of the generator. Therefore, the shunt braking method can be used as a centralized system for the wind farm in conjunction with a dynamic voltage restorer (DVR), which is discussed subsequently.

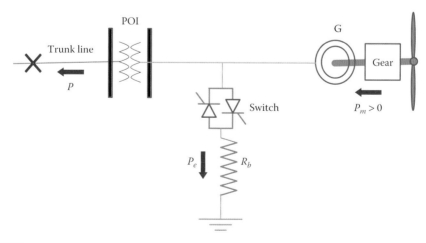

FIGURE 12.27
Shunt braking resistance circuit.

12.2.3.3 Dynamic Voltage Restorer

DVR was originally developed to protect sensitive load, such as semiconductor fabrication plants, from voltage fluctuations in power systems. The same technology can be used to regulate the voltage at the wind farm during faults. The general configuration of the DVR is shown in Figure 12.28. Its main components are a transformer with the high-voltage side connected in series with the generator, voltage source converter, and energy storage device. Because of the series transformer, the voltage across the turbine is

$$\bar{V}_{a1} = \bar{V}_{poi} + \bar{V}_{DVR} \tag{12.30}$$

where:
$\bar{V}_{a1}$ is the terminal voltage of the generator
$\bar{V}_{poi}$ is the grid voltage at the point of interconnection
$\bar{V}_{DVR}$ is the voltage across the high-voltage winding of the transformer

The voltage V_{DVR} is

$$V_{DVR} = \frac{N_h}{N_l} V_i \tag{12.31}$$

where:
N_h is the number of turns of the high-voltage winding
N_l is the number of turns of the low-voltage winding
V_i is the output of the dc/ac converter

As seen in Chapter 5, the dc/ac converter can produce output voltage with adjustable magnitude and phase shift. Hence, the magnitude and phase shift of V_i (and V_{DVR}) are controllable. Under normal operation, the output voltage of the dc/ac converter is zero, hence $V_{DVR} = 0$. When the voltage at the point of interconnection is reduced due to faults, the DVR produces a voltage with specific magnitude and angle to maintain the voltage $\bar{V}_{a1}$ at the generator side constant, as shown in the phasor diagrams in Figure 12.29.

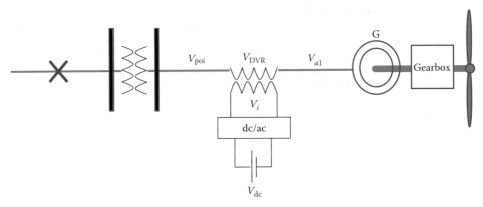

FIGURE 12.28
Dynamic voltage restorer.

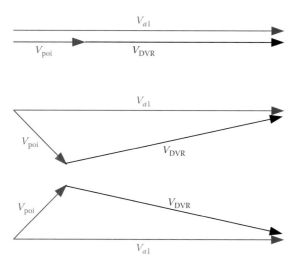

FIGURE 12.29
DVR under stressed conditions.

EXAMPLE 12.7

A 690 V WT is connected to the POI through a DVR. A fault caused the voltage at the POI to drop to $100\ \angle 30°\,\text{V}$. Compute the DVR voltage that maintains the voltage of the generator at 600 V.

Solution:
Using Equation 12.30, we get

$$\bar{V}_{\text{DVR}} = \bar{V}_{a1} - \bar{V}_{\text{poi}} = \frac{600}{\sqrt{3}}\ \angle 0° - \frac{100}{\sqrt{3}}\ \angle 30° = 297.81\ \angle 5.56°\,\text{V}$$

The DVR can be installed at the individual turbine or at a centralized location for the entire WPP, which is more cost effective. The DVR can be used in conjunction with the shunt braking method, as shown in Figure 12.30. Because DVR does not allow V_{a1} to collapse during faults, the shunt dynamic braking method can be effective in riding through faults.

EXAMPLE 12.8

For the system in Example 12.7, compute the shunt braking resistance that consumes 1 MJ in 2 seconds.

Solution:

$$P = \frac{E}{t} = \frac{10^6}{2} = 5 \times 10^5\,\text{W}$$

$$R_b = \frac{V_{a1}^2}{P} = 3 \times \frac{297.81^2}{5 \times 10^5} = 0.532\,\Omega$$

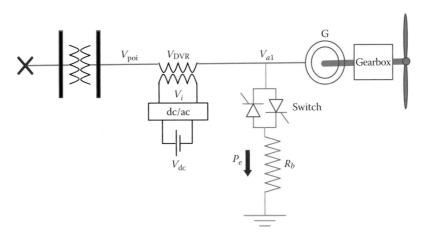

FIGURE 12.30
DVR with shunt braking resistance.

Beside the LVRT capability, the DVR can be used in normal operation to achieve two important functions:

Energy storage: The battery can absorb or provide real power. Thus, it can be used for power regulation.

Reactive power control: The voltage V_i can be adjusted to inject or consume reactive power. This feature is explained in Section 12.4.2.4.

12.3 Variability of the Wind Power Production

The variability and unpredictability of wind speed are among the most difficult problems facing wind energy integration. These problems are challenging to utility engineers as they have less control on the energy generated by from wind.

In addition to the variability of wind speed, wind energy production often peaks at night when the demand is low. This phenomenon has been observed worldwide. If the local laws prevent the WPP from shutting down when wind is available, the energy from wind is sold at reduced price or, in some cases, at negative price (WPP pays to sell its energy).

12.3.1 Uncertainty of Wind Speed

Wind speed forecasting is normally divided into two categories:

1. *Synoptic scale meteorology:* It predicts air mass, fronts, and pressure systems and is suitable for regional wind speed prediction. The synoptic forecast is reasonably accurate, but does not provide useful information for a small specific part of land.

2. *Mesoscale meteorology:* It predicts the effects of topography, bodies of water, and urban heat island on local wind speed at low elevations which is the needed forecast for wind farms.

The mesoscale wind speed forecasting is loosely divided into the following four types:

1. *Very short term:* Minutes ahead forecasting. It is also known as real-time wind forecasting.
2. *Short term (ST):* Hours ahead forecasting.
3. *Medium term (MT):* A day ahead forecasting.
4. *Long term (LT):* A week ahead forecasting.

To date, there is no accurate LT wind speed forecasting available. For the ST (0–6 h), the forecast is reasonably accurate for weather-stable regions.

The challenges facing wind forecasting include the following:

- Regional data cannot be used to forecast local conditions with specific terrains.
- Effects of local topology are hard to consider.
- Sufficient local sampling is needed, which requires years of monitoring at the WPP's site.
- Knowledge of features that impact wind speed is not fully known for every specific site.

Several forecasting techniques, including the standard regression methods, neural networks, and heuristic forecasting, are proposed for wind energy systems. Until the accuracy of the ST and MT forecasting increase, utility operates the power system in a more stochastic way. This requires the utility to change its operating algorithms and equipment to handle more variable generation. One method is to assume that wind power is a negative load. Hence, error in wind power forecasting is added to the error in load (demand) forecasting to produce a smaller error. This is because wind speed and load consumption are often orthogonal variables since electrical loads are not highly dependent on wind speed. Because of this orthogonality, the total error is less than the error of wind power forecasting plus the error in load forecasting. To explain this mathematically, assume that the load forecasting at a given utility is

$$P_{df} = P_d + \varepsilon_d \tag{12.32}$$

where:
P_d is the actual demand
P_{df} is the forecasted demand
ε_d is the error in demand forecast

For the wind power forecast

$$P_{wf} = P_w + \varepsilon_w \tag{12.33}$$

where:
P_w is the actual wind power
P_{wf} is the forecasted wind power
ε_w is the error in wind power forecast

Because ε_d and ε_w are often statistically independent (orthogonal), the total error ε is

$$\varepsilon \approx \sqrt{\varepsilon_d^2 + \varepsilon_w^2} \qquad (12.34)$$

Hence,

$$\varepsilon < (\varepsilon_d + \varepsilon_w) \qquad (12.35)$$

Because the total error is less than the sum of the two errors, the grid operator needs a smaller amount of reserves to balance the net load (load plus wind). Higher reserves would be needed if the operator balances the output of WPP in isolation of the load.

EXAMPLE 12.9

At a given time, the demand forecast is 2 GW with error of 2%. The forecasted wind power at the same time is 100 MW with forecasting error of 30%. Compute the total forecasting error.

Solution:
Compute the errors

$$\varepsilon_d = 2000 \times 0.02 = 40\,\text{MW}$$

$$\varepsilon_w = 100 \times 0.3 = 30\,\text{MW}$$

Total error

$$\varepsilon = \sqrt{\varepsilon_d^2 + \varepsilon_w^2} = \sqrt{40^2 + 30^2} = 50\,\text{MW}$$

The reserve needed to balance the power is 50 MW.
    If the output of WPP is balanced in isolation of the load, the total reserve is

$$\varepsilon_d + \varepsilon_w = 40 + 30 = 70\,\text{MW}$$

12.3.2 Variability of Wind Power Output

Wind energy installations are either clustered or distributed throughout the power systems. Clustered generation (wind farms or WPPs) is popular in the United States, where a large number of WTs are placed in one location. The distributed generation, also known as "wind garden," is mainly a European model, where small numbers of WTs are installed at distributed sites over a wide geographical area.
    The advantages of the clustered model include the following:

- Effectively use of areas with high-quality wind resources.
- WTs are installed much more quickly because the expensive and heavy equipment needed to install the turbines are not transported from one site to the other.
- Control and maintenance of WTs are cost effective.
- Developers deal with only a single or small number of landowners.
- Only one set of local regulation need to be satisfied.

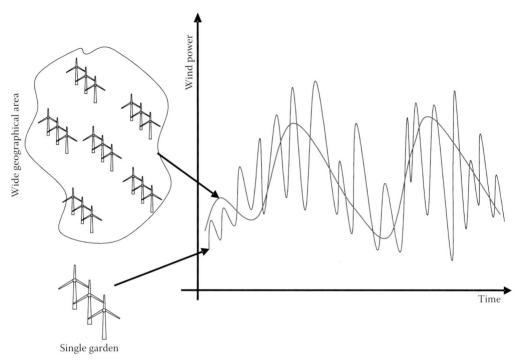

FIGURE 12.31
Wind generation output in distributed system.

The cluster model, however, is highly susceptible to shifting wind speeds. Because WPP often has large power production, the grid is more sensitive to variations in wind conditions and wind speeds. Key utility operational functions, such as unit commitment, scheduling, and dispatch, are harder to implement.

The distributed model is more expensive as the turbines are installed in various geographical areas. However, each wind garden is too small to have an impact on grid operation. Because wind shifts are not expected to be synchronized at all gardens, wide variation of wind generation is reduced. The output of all wind gardens in a wide geographical area can be easier to predict than that for WPP model, and the scheduling over the wide area is much easier. Figure 12.31 depicts the normalized output of a single-wind garden versus distributed wind gardens. A single-wind garden is expected to have wide variations in its output based on its local wind conditions. If several of these gardens are placed over a wide geographical area (such as a nation), smoother variations in generation from all wind gardens is expected.

12.3.3 Balancing Wind Energy

Because of the continuous change in wind speed, electric power generated by WT fluctuates on all time scales. This gives power system operators less control on balancing the energy of the grid. To ramp down generation from wind is the easier problem, as it requires spilling more wind. However, ramping-up generation is more difficult as we have no control on wind speed. Among the methods suggested for the ramp up of WPP are the following:

Scale back generation: WPP reduces its generation below the available power for the given wind condition. This way when wind speed is reduced, pitch control adjusts production to the earlier level. This function is the same as the delta control discussed in Section 12.1.3.

Local spinning reserve: WPP installs balancing generation (mainly gas turbines) on site that are used to ramp up generation when needed.

Balancing service: WPP contracts with a separate utility to provide the needed extra energy when wind speed is reduced. The utility responsible for balancing generation to load are called balancing areas. The United States has about 130 balancing areas. The largest one is the PJM grid with a peak of 145 GW in 2013.

In addition to the above methods, storage and load management can counterbalance the variation of wind energy production. These two methods are discussed subsequently.

12.3.3.1 Energy Storage

Energy storage is one of the most crucial aspects of renewable energy and smart grid. It allows utilities to better integrate renewable systems by implementing load following, peak load shaving, voltage support, oscillations damping, frequency regulation, and ride-through faults. In addition, energy storage reduces the need for additional transmission assets, improves the reliability of electricity supply, increases the efficiency of existing power plant and transmission facilities, reduces the investment required for new facilities, and reduces energy cost.

Energy storage systems include batteries, pump hydro, compressed air, flywheels, super capacitors, superconductor magnetic storage, thermal storage, and hydrogen. Some of these technologies are well developed, but most are still expensive or not yet reliable enough for grid implementation.

Battery is one of the oldest electric energy storage methods; it was invented by Volta in the eighteenth century. The main problems associated with most batteries include their low energy capacity/volume, slow charging, and short lifespan. Recently, a new generation of batteries, which is much better than its predecessors, has been developed. Most of them have energy densities between 100 and 500 KJ/Kg, and their single-cell voltage is between 1.2 and 4.2 V. On the low end of energy density are the lead–acid, which are commonly used in automobiles and nickel–cadmium batteries. Their voltages are roughly 2.1 and 1.2 V per cell, respectively. Their energy density is about 140 kJ/kg. On the high end of energy density and voltage is the lithium-ion battery (3.6 V, 500 kJ/kg). Although expensive, it has a very low rate of self-discharge. One drawback of this battery is its ability to overheat, especially during short circuits. Lithium-ion battery systems in the MW range are already installed throughout power grids. The lithium-ion battery is also used in electric vehicles and aviation. Sodium–sulfur is another type of the high-density batteries (2 V, 500 kJ/kg). The drawback of this battery is the spontaneous burns of sodium when exposed to air and moisture. Moreover, when corroded, its self-discharge rate increases. Nickel–metal hydride battery is also a high-density battery, but at low voltage (1.2 V, 360 kJ/kg). It is relatively inexpensive, but has a high rate of self-discharge. Nickel–zinc battery is similar to the nickel–metal hydride, but has a lower self-discharge rate.

For WPP, energy storage can be used in mainly two applications: energy regulation and energy arbitrage. Storage system can compensate for wind fluctuations as well as demand variations. In addition, as the energy from WPP during low demands is not cost effective, the energy can be stored and sold at favorable rate during periods of higher demand.

12.3.3.2 Load Management

Electric power system is designed and operated to meet load demand at all times. The balance of energy is almost always done at the generation side, and there is no control on the load side. In the United States, only 1% of peak system load is controllable. With today's technology, it is possible to control the energy consumption of some equipment without inconveniencing customers. For example, if appliances such as water heater or deep freezer are turned off for a few minutes, the user may not notice any change. But if enough of these actions are implemented at various times, they can help the grid balance its energy production to demands even if part of the production is stochastic (wind, solar, etc.).

If we look broadly at the energy picture in the United States, we find that about 18% of the total energy is consumed by household appliances that can be controlled such as air conditioners, refrigerators, water heaters, stoves, washers, and dryers. Controlling these appliances can increase the spinning reserve nationwide from 13% to over 30%. In addition, the industrial and commercial sectors consume roughly 56% of the total energy. Some of their loads can be controlled as well.

12.4 Reactive Power

WPPs are required to provide reactive power support to the grid similar to the requirements from conventional power plants. As explained in Chapters 10 and 11, types 3 and 4 systems can adjust their reactive power output through their pulse width modulation (PWM) techniques. However, types 1 and 2 systems cannot generate reactive power on their own and their reactive power needs come from the grid. If the reactive power control of WPP turbines is not sufficient to meet interconnection requirement, external devices can be used such as the static VAR compensators (SVC) and the synchronous condenser (SC).

12.4.1 Turbine Reactive Power Control

In Chapters 10 and 11, the reactive power of types 3 and 4 systems are evaluated. One of the great advantages of these systems is their ability to control the power factor at the point of interconnection even when the system voltage is substantially reduced during faults. In either type, the grid-side converter (GSC) is a dc/ac converter, as shown at the top part of Figure 12.32. The voltage of the dc bus (V_{dc}) is converted into an ac voltage (V_{out}). The magnitude and phase shift of V_{out} can be adjusted by the PWM technique discussed in Chapter 5. The converter is connected to the POI through transformer and perhaps short line. The system can be modeled as shown at the bottom part of Figure 12.32, where

$$\overline{V}'_{out} = \frac{N_1}{N_2}\overline{V}_{out} = \overline{V}_{poi} + \overline{I}\overline{x}$$

$$x = x_{line} + x_{xfm}$$

(12.36)

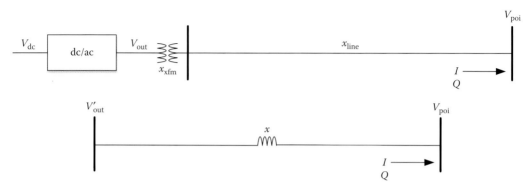

FIGURE 12.32
Grid-side converter connected to POI.

where:

N_1 is the number of turns of the high-voltage winding of the transformer (grid-side winding)

N_2 is the number of turns of the low-voltage windings

V'_{out} is V_{out} referred to the high-voltage side of the transformer

x_{line} is the inductive reactance of the line

x_{xfm} is the inductive reactance of the transformer

The reactive power delivered to the POI is

$$Q = Imag\left(\bar{V}_{poi}\,\bar{I}^{*}\right) \tag{12.37}$$

where:

$$\bar{I} = \frac{\bar{V}'_{out} - \bar{V}_{poi}}{jx} \tag{12.38}$$

Taking $\bar{V}_{poi}$ as the reference phasor and substituting the current in Equation 12.38 into Equation 12.37 yields

$$Q = \frac{V_{poi}}{x}\left(V'_{out}\cos\delta - V_{poi}\right) \tag{12.39}$$

where:

δ is the angle of $\bar{V}'_{out}$, which is controlled by the PWM circuit

Equation 12.39 shows that the reactive power delivered to the POI is positive when V_{out} is adjusted, so that ($V'_{out}\cos\delta > V_{poi}$). Positive reactive power means that the generator is delivering reactive power to the POI.

Even when the generator is producing no real power, reactive power can still be delivered to the POI to support system voltage. In this case, the power angle δ is set to zero; V'_{out} is in phase with V_{poi}. Hence, $\cos\delta = 1$, and the reactive power delivered at the POI is

$$Q = \frac{V_{poi}}{x}\left(V'_{out} - V_{poi}\right) \tag{12.40}$$

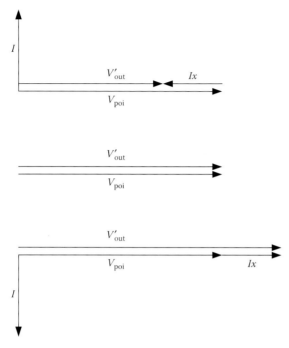

FIGURE 12.33
Phasor diagram of the system in Figure 12.32 when no real power is delivered.

When no real power is produced by the wind farm, the phasor diagrams representing Equation 12.36 is shown in Figure 12.33. Based on the magnitude of V'_{out} with respect to V_{poi} the reactive power at the POI can be one of three statuses:

- In the top phasor diagram, the output of the converter is adjusted, so that $V'_{out} < V_{poi}$. In this case, the current is leading (the current always lags phasor Ix by 90°); meaning that the POI is seen by the converter as a capacitive load. Q, in this case, is negative.
- In the middle phasor diagram, the output of the converter is adjusted, so that $V'_{out} = V_{poi}$. In this case, no current is flowing in the circuit. Thus, no reactive power is at the POI.
- In the bottom phasor diagram, the output of the converter is adjusted, so that $V'_{out} > V_{poi}$. In this case, the current is lagging and the POI is seen as inductive load-consuming reactive power. Q, in this case, is positive.

EXAMPLE 12.10

The PWM circuit of the GSC is adjusted to produce output voltage of 700 V (line-to-line) that is in phase with the POI bus voltage. The transformer between the GSC and the transmission line is Y–Y connected with a turns ratio of 22 and zero phase shift. The equivalent inductive reactance of the transformer as referred to the high-voltage side is 4 Ω. The inductive reactance of the transmission line between the converter and the POI is 2 Ω. If the POI voltage is 15.0 kV, compute reactive powers delivered to the grid.

Solution:

The first step is to refer the output voltage of the full converter to the high-voltage side of the transmission line.

$$V'_{out} = V_{out} \times 22 = 700 \times 22 = 15.4\,\text{kV}$$

The total inductive reactance between the full converter and the POI bus is

$$x = x_{xfm} + x_{line} = 4 + 2 = 6\,\Omega$$

The reactive power at the POI is

$$Q = \frac{V_{poi}}{x}(V'_{out} - V_{poi}) = 3 \times \frac{\left(15/\sqrt{3}\right)}{6}\left(\frac{15.4}{\sqrt{3}} - \frac{15}{\sqrt{3}}\right) = 1.0\,\text{MVAR}$$

EXAMPLE 12.11

For the system in Example 12.10, compute V_{out} that delivers 600 kVAr to the POI during grid fault that reduces V_{poi} to 3 kV (line-to-line). Assume that there is no real power generated by the turbine.

Solution:

Use the reactive power equation to compute V'_{out}.

$$Q = \frac{V_{poi}}{x}(V'_{out} - V_{poi})$$

$$\frac{0.6}{3} = \frac{\left(3/\sqrt{3}\right)}{6}\left(V'_{out} - \frac{3}{\sqrt{3}}\right)$$

$$V'_{out} = 2.425\,\text{kV}$$

Hence, the line to line voltage of V_{out} is

$$V_{out} = \sqrt{3}\,\frac{V'_{out}}{22} = 190.91\,\text{V}$$

12.4.2 Static VAR Compensator

Static VAR compensator (SVC) is a term given to a family of solid-state devices designed to regulate reactive power at a rapid rate. These devices are also capable of other applications such as stabilizing voltage, improving system stability, and correcting phase unbalance. The first SVC installation was in 1977 at Basin Electric Power Cooperative in Nebraska. Since then, these devices became quite popular and are extensively used all over the power grids.

There are five main types of SVC systems, which are as follows:

1. Thyristor-controlled reactor (TCR)
2. Thyristor-switched capacitor (TSC)
3. TSR-TSC
4. Static compensator (STATCOM)
5. DVR

12.4.2.1 Thyristor-Controlled Reactor

The key components of the TCR are shown in Figure 12.34. It consists of a step-down transformer, an air-core inductor, and a solid-state anti-parallel switch. The air-core inductor is chosen to avoid core saturation that reduces the value of the inductive reactance during heavy currents. The transformer is used to reduce the voltage ratings across the inductor and solid-state switches.

Because the natural zero crossing of an inductive current occurs when the voltage is at its peak (the inductive current lags the voltage by 90°), the triggering angle of the forward silicon controlled rectifier (SCR) is greater or equal to 90°, and the reverse SCR is triggered after an additional 180°, as shown in Figure 12.35. The voltage waveform is for the low-voltage side of the transformer, and the current is for the inductor current. During the positive half of the voltage cycle, if the triggering angle $\alpha = 90°$, the current is continuous. If $\alpha > 90°$, the current flows until the commutation angle β. What determines

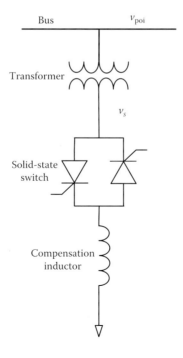

FIGURE 12.34
Thyristor-controlled reactor.

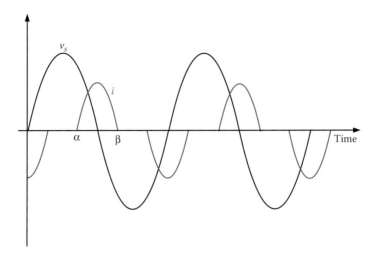

FIGURE 12.35
Waveforms of TCR.

β is the amount of inductor energy that is acquired then dissipated during one switching cycle of each SCR. Because inductors do not permanently store energy, the average voltage across the inductor (V_{ave-L}) is zero.

$$V_{ave-L} = \frac{1}{\pi} \int_{\alpha}^{\beta} v_s \, d\omega t = \frac{1}{\pi} \int_{\alpha}^{\beta} V_{max} \sin \omega t \, d\omega t = \frac{V_{max}}{\pi} (\cos \alpha - \cos \beta) = 0 \qquad (12.41)$$

Hence,

$$\beta = 2\pi - \alpha \qquad (12.42)$$

The conduction period γ is defined as

$$\gamma = \beta - \alpha = 2(\pi - \alpha) \qquad (12.43)$$

The rms voltage across the inductor (V_{rms-L}) is

$$V_{rms-L} = \sqrt{\frac{1}{\pi} \int_{\alpha}^{\beta} v_s^2 \, d\omega t} = \frac{V_{max}}{\sqrt{2\pi}} \sqrt{(\gamma + \sin 2\alpha)} \qquad (12.44)$$

The inductive power consumed by the TCR is

$$Q_L = \frac{V_{rms-L}^2}{\omega L} = \frac{V_{max}^2}{2\pi \omega L} [\gamma + \sin 2\alpha] = \frac{V_{max}^2/2}{\pi x_L} [\gamma + \sin 2\alpha]$$

$$\qquad (12.45)$$

$$Q_L = \frac{V_{rms-s}^2}{\pi x_L} [2(\pi - \alpha) + \sin 2\alpha]$$

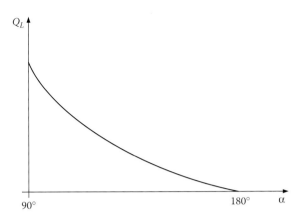

FIGURE 12.36
Reactive power of TCR.

where:
Q_L is the inductive power consumed by the TCR
L is the inductance
x_L is the inductive reactance
V_{rms-s} is the rms voltage of the source (bus voltage)

Figure 12.36 shows the reactive power consumed by the TCR. The maximum consumption occurs when $\alpha = 90°$. At $\alpha \geq 180°$, no reactive power is consumed.

EXAMPLE 12.12

The voltage on the low-voltage side of a TCR transformer is 690 V. The TCR inductive reactance is 50 mH. Compute the maximum reactive power of the TCR. Also, compute the triggering angle that would allow the TCR to consume 50% of its maximum reactive power.

Solution:
The maximum reactive power of the TCR can be computed by setting $\alpha = 90°$ in Equation 12.45

$$Q_L = \frac{V^2_{rms-s}}{\pi x_L}[2(\pi - \alpha) + \sin 2\alpha] = \frac{V^2_{rms-s}}{x_L} = \frac{690^2}{377 \times 0.05} = 25.257 \, \text{kVAr}$$

For 50% consumption

$$Q_L = \frac{V^2_{rms-s}}{\pi x_L}[2(\pi - \alpha) + \sin 2\alpha] = \frac{25257}{2}$$

$$\frac{690^2}{377 \times 0.05\pi}[2(\pi - \alpha) + \sin 2\alpha] = \frac{25257}{2}$$

By iteration

$$\alpha \approx 113.83°$$

Because the TCR is essentially a regulated inductor, it is used to consume surplus reactive power. For WPP, TCR is used for offshore systems with long cables to shore stations. Cables have high parasitic capacitance that could cause the voltage at the shore station to be higher than desired. In this case, the excess of reactive power from the cable capacitance is consumed by the TCR to bring the voltage to the desired level.

12.4.2.1.1 TCR-Fixed Capacitor

In more general applications, where the reactive power need to be either consumed or delivered to the grid, the TSC is combined with a fixed capacitor, as shown in Figure 12.37. The total reactive power Q_{total} in this case is

$$Q_{total} = Q_L - Q_c = \frac{V_{rms-s}^2}{\pi x_L}[2(\pi - \alpha) + \sin 2\alpha] - Q_c \qquad (12.46)$$

In this system, the reactor is sized higher in reactive power than that of the capacitor to allow for leading and lagging compensation. The total reactive power as a function of the triggering angle is shown in Figure 12.38. At α_0, the total reactive power is zero. When $\alpha < \alpha_0$, the TCR consumes reactive power, and when $\alpha > \alpha_0$, it is delivering reactive power.

For weak bus, the TSC with fixed capacitor can be used to maintain the bus voltage to a desired value. This is shown in Figure 12.39, where V_r is assumed to be the desired voltage. If the actual voltage, say V_1, is lower than V_r, the triggering angle increases above α_0 to produce more reactive power. If the actual voltage, say V_2, is higher than V_r, the triggering angle decreases below α_0 to consume more reactive power.

The main drawback of this SVC system is the continuous circulating current between the capacitor and inductor at all times. Even at zero reactive power, the TCR is switched to compensate for the capacitor reactive power. This drawback results in higher system losses (typically 0.5%–0.7%), and reduced the lifetime of its components.

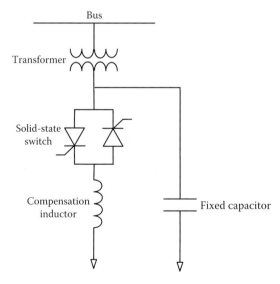

FIGURE 12.37
TCR with fixed capacitors.

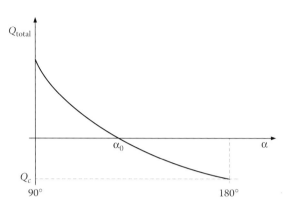

FIGURE 12.38
Reactive power of TCR with fixed capacitors.

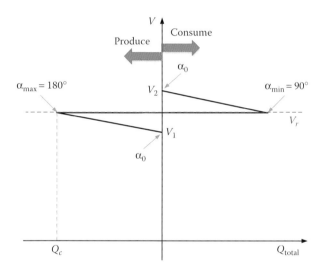

FIGURE 12.39
Voltage control of TCR with fixed capacitors.

12.4.2.1.2 TCR with Mechanically Switched Capacitor

To address the drawback of the TSC with fixed capacitor, the capacitors are mechanically switched when needed, as shown in Figure 12.40. Because mechanically switching capacitors produce damaging high-frequency transients, the capacitors are switched just a few times per day to reduce the impact of the damaging transients. In addition, a high-frequency filter is often used to absorb these transients.

Because the capacitors are switched when needed, this SVC system has lower losses than that for fixed capacitors (0.02%–0.05%). However, because of the low switching rate of the capacitors, the device cannot be used for fast adaptive compensation. Keep in mind that most mechanical switches require at least two cycles to close and eight cycles to open. In addition, capacitors have typically long discharging time as their built-in resistors take as much as 10 minutes to fully discharge the capacitors.

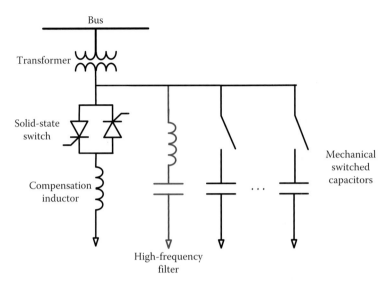

FIGURE 12.40
TCR with mechanical switched capacitors.

12.4.2.2 Thyristor-Switched Capacitor

TSC is designed to produce reactive power that can be adjusted every half cycle with minimum switching transients. The main components of the system are shown in Figure 12.41. It consists of capacitors that are switched by solid-state devices. To eliminate the switching transients, each SCR is closed when the voltage across it is zero. This is known as "zero-voltage-switching." Hence, the TSC can produce two different values of reactive power in one ac cycle. In series with the capacitor, a small snubbing inductor is used to reduce the transients in current when the switching of the SCRs is not precise. The capacitors of the TSC can be sized in a binary ratio to reduce the step size of the compensation.

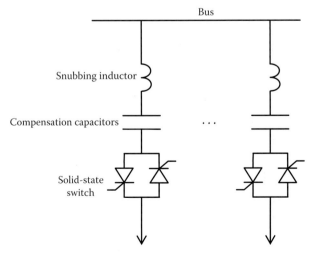

FIGURE 12.41
Thyristor-switched capacitor.

12.4.2.3 TSR-TSC

As seen in the earlier sections, the TSC produces reactive power and cannot consume reactive power. On the other hand, the TCR consumes reactive power and cannot produce reactive power. Thus, if we use both systems, we can have bidirectional compensation. This is the TSR-TSC system shown in Figure 12.42. It consists of several TSC banks and one TCR. The system is fast and can adaptively adjust the reactive power.

12.4.2.4 Static Compensator

The static compensator (STATCOM) is a highly versatile device that is fast acting and highly adaptive. Its main components are shown in Figure 12.43. It consists of a source of energy (battery or charged capacitor) connected to the grid through a dc/ac converter and a transformer. The dc/ac converter uses PWM technique to control the magnitude and phase shift of the output voltage of the converter injected into the transformer (V_i).

The equivalent circuit of the STATCOM is shown in Figure 12.44, where V_i' is the output voltage of the dc/ac converter as referred to the POI side. Because the flow of current is assumed from the STATCOM to the POI, the equation of the system is

$$\overline{V}_i' = \overline{V}_{poi} + jx_{xfm}\overline{I} \tag{12.47}$$

The equation of the reactive power delivered to the POI is similar to that in Equation 12.40.

$$Q = \frac{V_{poi}}{x}\left(V_i' - V_{poi}\right) \tag{12.48}$$

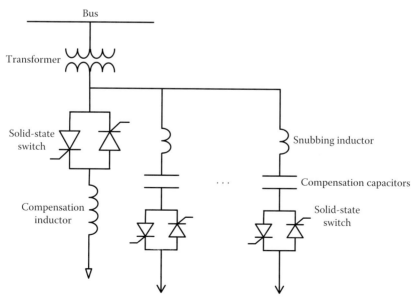

FIGURE 12.42
TCR with TSC.

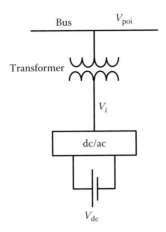

FIGURE 12.43
STATCOM.

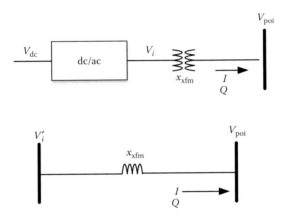

FIGURE 12.44
Equivalent circuit of STATCOM.

If V_i is adjusted so that V_i' is in phase with $\overline{V}_{poi}$, we can control the flow of reactive power delivered to the grid. Take, for example, the case in Figure 12.45 where V_i' is adjusted to be greater than V_{poi}. The current from the STATCOM is lagging and the device is delivering reactive power to the POI.

For the case in Figure 12.46, V_t' is adjusted to be smaller than V_{poi}. The current from the STATCOM is leading and the device is consuming reactive power from the POI bus.

12.4.3 Synchronous Condenser

The SC is one of the oldest methods of adaptive reactive power control. It was used extensively before the invention of the SVCs. It is still in use today, although at smaller scale. The device is just a cylindrical rotor synchronous machine running as a motor, and often under no load condition.

The model for the synchronous generator is given in Chapter 7. The model can be modified for the reverse flow of current when the machine is running as a motor. Because the

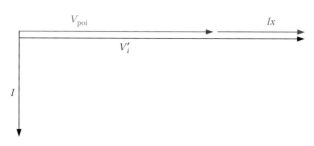

FIGURE 12.45
STATCOM delivering reactive power at POI.

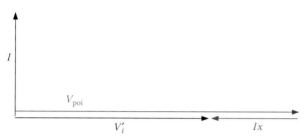

FIGURE 12.46
STATCOM consuming reactive power at POI.

current is going into the machine, the machine is the load. Hence, negative reactive power means delivery and positive means consumption.

$$Q = 3\frac{V_a}{x_s}\left(V_a - E_f \cos\delta\right) \qquad (12.49)$$

where:
  V_a is the terminal voltage of the synchronous machine
  E_f is the equivalent excitation voltage of the machine (it is a function of the field current)
  x_s is the synchronous reactance of the machine
  δ is the power angle (angle between E_f and V_a)

If the machine is running at no load, the real power consumed is just enough to compensate for the losses. If we ignore these losses, the power angle at no load is zero. Hence, Equation 12.49 is

$$Q = 3\frac{V_a}{x_s}\left(V_a - E_f\right) \qquad (12.50)$$

If the excitation is adjusted so that $E_f > V_a$, the reactive power is negative meaning the machine is delivering reactive power. If $E_f < V_a$, the reactive power is positive meaning the machine is consuming reactive power.

Exercise

1. State five integration challenges for wind power plants.
2. What causes the power angle of the synchronous machine to change?
3. What is the role of frequency in the change of power angle?
4. What is power system stability?
5. What is power system security?
6. Can a power system be stable and insecure?
7. What is the pullout power?
8. How to increase the pullout power of a synchronous generator?
9. How to increase the pullout power of an induction generator?
10. What is delta control?
11. Why is it important to have wind turbine fault ride-through?
12. What is the impact of faults on drive trains?
13. What is the low-voltage ride-through requirement?
14. Under what conditions is the turbine allowed to disconnect from the grid during faults?
15. What are the ramping requirements during contingencies?
16. What are the ramping requirements during steady state operation?
17. What is the function of dynamic braking?
18. How effective is the shunt dynamic braking during bolted faults?
19. What is the main objective of a DVR?
20. What is mesoscale forecasting?
21. What are the problems associated with the variability of wind speed?
22. State four methods to balance the energy due to the variability of wind speed.
23. Give an example of load management.
24. How is reactive power controlled by a type 3 turbine?
25. How is reactive power controlled by a type 4 turbine?
26. How is reactive power controlled by wind farms?
27. What are the advantages and disadvantages of TCR?
28. What are the advantages and disadvantages of TSC?
29. What are the advantages and disadvantages of TCR-TSR?
30. What are the advantages and disadvantages of STATCOM?
31. What are the advantages and disadvantages of synchronous condensers?
32. A synchronous generator of a type 4 wind turbine is connected to the grid through a transmission line of 0.5 pu inductive reactance. The transformer of the full converter has an inductive reactance of 0.5 pu. The voltage of the grid is 1.0 pu and the output voltage of the full converter is 1.0 pu. Can the turbine deliver 1.2 pu real power?

33. A synchronous generator of a type 4 wind turbine is connected to the grid through a transmission line of 0.2 pu inductive reactance. The transformer of the full converter has an inductive reactance of 0.3 pu. The voltage of the grid is 1.0 pu. The output voltage of the full converter is adjusted to 1.0 $\angle 30°$. Compute the real and reactive power delivered to the POI.

34. A synchronous generator of a type 4 wind turbine is connected to the grid through a transmission line of 0.2 pu inductive reactance. The transformer of the full converter has an inductive reactance of 0.3 pu. The voltage of the grid is 1.0 pu. The power angle is adjusted to 20°. Compute the output voltage of the converter that delivers 0.5 pu reactive power to the grid. Compute the real power delivered to the grid under this condition.

35. A 700 V, Y-connected, type 4 synchronous generator is connected to the POI through a trunk line and transformer. The synchronous reactance of the generator as referred to the high-voltage side of the transformer is 4.0 Ω The inductive reactance of the trunk line plus transformer as referred to the high-voltage side is 6.0 Ω. The turns ratio of the transformer is 20 and the grid POI voltage is 13.8 kV (line-to-line). A grid fault occurred that reduces the POI voltage to 2 kV and the delivered power to 0.5 MW. Compute the steady-state current feeding the fault. Also, compute the reactive power at the POI.

36. A type 2 wind turbine has a six-pole, 60 Hz three-phase, Y-connected induction generator. The terminal voltage of the generator is 690 V. The parameters of the machine are $r_2' = 10$ mΩ and $x_{eq} = 0.1$ Ω.

    A 0.02 Ω resistance is inserted into the rotor circuit (as referred to the stator) through a solid-state switching circuit with 50% duty ratio. Ignore all losses. Can this system deliver 3.1 MW to the grid?

37. A 690 V, six-pole, Y-connected, type 3 generator has the following parameters: $r_2' = 0.01$; $x_1 = x_2' = 100$ mΩ and $x_m = 5$ Ω.

    The inductive reactance of the trunk line plus transformer as referred to the low-voltage side is 1.0 Ω. Compute the injected voltage that limits the stator current to 250 A during bolted fault.

38. A 690 V wind turbine was operating normally when a bolted fault occurred at the POI. Size the braking resistance that limits the current of the generator to 1.0 kA and absorb the energy in 2 seconds.

39. A 690 V type 3 wind turbine is connected to the POI through a DVR and a transformer of 1 Ω inductive reactance. A bolted fault occurred at the POI. Compute the DVR voltage that maintains the voltage of the generator at 690 V and limit the fault current to 200 A.

40. The forecasted average demand of a given utility is 5.0 ± 2% GW. The forecasted average wind power at the same utility for the same period is 200 ± 25% MW. Compute the reserve needed to meet the average load demand.

41. The PWM circuit of the GSC of a type 4 system adjusts its output voltage to 800 V (line-to-line) with 2° lagging angle with respect to the POI voltage. The transformer between the GSC and the transmission line is Y-Y connected with a turns ratio of 20. The equivalent inductive reactance of the transformer plus the transmission line between the converter and the POI is 6 Ω (referred to the high-voltage

side of the transformer). If the POI voltage is 15.0 kV, compute the real and reactive powers delivered to the grid.

42. The voltage on the low-voltage side of a TCR transformer is 690 V. The TCR inductive reactance is 50 mH. Compute the triggering angle that would allow the TCR to consume 50% of its maximum reactive power.

Index

Note: Locators followed by "*f*" and "*t*" denote figures and tables in the text